Lecture N
Mathemat

T0259750

Edited by A. Dold and B. Eckmann

Subseries: Instituto de Matemática Pura e Aplicada, Rio de Janeiro
Adviser: C. Camacho

1050

Alexander Prestel
Peter Roquette

Formally p-adic Fields

Springer-Verlag
Berlin Heidelberg New York Tokyo 1984

Authors

Alexander Prestel
Fakultät für Mathematik, Universität Konstanz
Postfach 5560, 7750 Konstanz, Federal Republic of Germany

Peter Roquette
Mathematisches Institut, Universität Heidelberg
Im Neuenheimer Feld 288, 6900 Heidelberg, Federal Republic of Germany

This book is also available as no. 38 of the series "Monografias de Matemática", published by the Instituto de Matemática Pura e Aplicada, Rio de Janeiro

AMS Subject Classifications (1980): 12B99, 12J10

ISBN 3-540-12890-5 Springer-Verlag Berlin Heidelberg New York Tokyo
ISBN 0-387-12890-5 Springer-Verlag New York Heidelberg Berlin Tokyo

This work is subject to copyright. All rights are reserved, whether the whole or part of the material is concerned, specifically those of translation, reprinting, re-use of illustrations, broadcasting, reproduction by photocopying machine or similar means, and storage in data banks. Under § 54 of the German Copyright Law where copies are made for other than private use, a fee is payable to "Verwertungsgesellschaft Wort", Munich.

© by Springer-Verlag Berlin Heidelberg 1984
Printed in Germany

Printing and binding: Beltz Offsetdruck, Hemsbach/Bergstr.
2146/3140-543210

Preface

These notes result from lectures given by the authors at IMPA in Rio de Janeiro - in 1980 by the second and in 1982 by the first author. The 1980 course was mainly concerned with the content of Sections 6 and 7 , using as a prerequesite the Ax-Kochen-Eršov- Theorem on the model completeness of the theory of p-adically closed fields. After that, an algebraic approach to this important theorem as well as an analysis of p-adic closures was developed (contained in Sections 3 and 4). The present lecture notes essentially coincide with the 1982 course.

In the introductory Section 1 we try to point out the analogy between the theory of p-valued fields and the well-known theory of ordered fields. After giving some basic definitions and examples as well as basic facts from general valuation theory in Section 2, we develop the theory of p-valued fields, i.e. fields together with a fixed p-valuation in Section 3 and 4 . From Section 6 on we no longer fix a certain p-valuation, instead we only assume the existence of such a valuation. Fields which admit some p-valuation are called formally p-adic. The theory of formally p-adic fields is concerned with the investigation of all p-valuations rather than just one. In Section 7 we concentrate on the important case of function fields. In Section 5 we use results proved in previous sections for model theoretic investigations of formally p-adic fields. In particular we deduce the Ax-Kochen-Eršov-Theorem. Viewed historically, these results stand at the beginning of the development of formally p-adic fields.

The only Section which makes use of model theoretic notions and facts is Section 5 . However, there is one exception - the notion of saturated structures - which is used also in the formulation and proof of the Embedding Theorem in Section 4 . We consider this notion as very useful in the investigation of function fields, since with its help 'specialization theorems' become essentially equivalent to 'embedding theorems'. Actually, it is this equivalence which is used in Section 7 , Theorem 7.2 . In order to keep this book as selfcontained as possible, we add in Proposition 7.3 elementary proofs of all the facts needed here about saturated fields. The reader who is interested in the general theory of saturated structures is referred to the books [B-S],[C-K], and [S].

The authors wish to thank all colleagues who offered helpful suggestions in the preparation of these notes, and owe a special debt to F.V. Kuhlmann for reading the complete manuscript and setting up the notation and the subject index. Last but not least we are grateful to Edda Polte for preparing the typescript.

Heidelberg-Konstanz 1983

Contents

§ 1. Introduction and motivation

"The notion of a p-adically closed field axiomatizes the algebraic properties of the p-adic number field $\mathbb{Q}_p$ in the same way as the notion of a real closed field axiomatizes the algebraic properties of the real number field $\mathbb{R}$".

In this introductory section we shall explain what we mean by this statement. While introducing the basic notions and mentioning the main theorems, we will always compare the 'p-adic' case with the 'real' case. In later sections we will mention the real case only occasionally. In order to keep this introduction as transparent as possible we will restrict ourselves here to the axiomatization of $\mathbb{Q}_p$, although , in the following sections we shall consider the more general situation of the completion of a number field K of finite degree over $\mathbb{Q}$ with respect to a fixed finite prime $\mathfrak{P}$ of K .

As it is well-known, $\mathbb{R}$ and $\mathbb{Q}_p$ (where p is any prime number from $\mathbb{N}$) are the completions of the rational number field $\mathbb{Q}$ with respect to the usual absolute value $|\,|$ defined by

$$|x| = \begin{cases} x & \text{if } 0 \leq x \\ -x & \text{if } x \leq 0 \end{cases}$$

and the p-adic absolute value $|\,|_p$ defined by

$$|x|_p = p^{-v_p(x)}$$

where the integer $v_p(x)$ is uniquely determined (in case $x \neq 0$) by $x = p^{v_p(x)} \cdot \frac{m}{n}$ with $m \in \mathbb{Z}$, $n \in \mathbb{N}$ and n,m prime to p .

The absolute values of $\mathbb{Q}$ have canonical extensions to the corresponding completions.

Since the development of the theory of p-adically closed fields is in almost complete analogy to that of the real closed fields, let us start with an exposition of the latter one. As a general reference about formally real fields let us mention $[P_1]$.

In his famous talk at the International Congress of Mathematicians in 1900 Hilbert proposed as his 17th problem to prove that every polynomial $f(X_1,\ldots,X_n) \in \mathbb{R}[X_1,\ldots,X_n]$ which is positive definite on $\mathbb{R}$ (i.e. $f(a_1,\ldots,a_n) \geq 0$ for all $a_1,\ldots,a_n \in \mathbb{R}$) equals a sum of squares of rational functions in $X_1,\ldots,X_n$ over $\mathbb{R}$. (Originally this problem came up in his foundation of geometry and was formulated for polynomials with rational coefficients; cf. $[P_2]$.) In 1926 Artin solved Hilbert's 17th Problem positively (see [A]). The basic notion used in his proof was the notion of an ordering.

An <u>ordering</u> on a field K is a binary relation $\leq$ satisfying the following conditions:

o_1 : $x \leq x$

o_2 : $x \leq y,\ y \leq z \Rightarrow x \leq z$

o_3 : $x \leq y$ or $y \leq x$

o_4 : $x \leq y \Rightarrow x + z \leq y + z$

o_5 : $0 \leq x,\ 0 \leq y \Rightarrow 0 \leq x \cdot y$

o_6 : $1 \nleq 0$

The axioms of an ordering constitute the most basic rules of computation for the usual ordering of the reals. The abstract notion of an ordering was already well-known at the time when Artin made use of it. However, only after his work it became clear that $o_1 - o_6$ are not just some basic rules for $\leq$ in $\mathbb{R}$. They form a complete set of axioms for all 'rules' of $\leq$ in $\mathbb{R}$, i.e. every other 'rule' is a logical consequence of $o_1 - o_6$. We will explain this further once we come to the p-adic analog.

Returning to Artin's work, let us list the two main theorems which lead to a positive solution of Hilbert's 17th Problem. The first one characterizes the <u>totally positive</u> elements of a field K, i.e. elements $a \in K$ which are positive with respect to every ordering of K.

THEOREM 1 <u>An element</u> $a \in K$ <u>is totally positive if and only if it is a sum of squares in</u> K.

The second theorem characterizes function fields of varieties V defined over $\mathbb{R}$ which have non-singular real points.

THEOREM 2 <u>Let</u> V <u>be an affine variety defined over</u> $\mathbb{R}$. <u>Then</u> V <u>admits a non-singular real point if and only if the function field</u> $\mathbb{R}(V)$ <u>admits an ordering</u>.

Now it is easy to solve Hilbert's 17th Problem: Let $f \in \mathbb{R}[X_1,\ldots,X_n]$ be different from every sum of squares in $K = \mathbb{R}(X_1,\ldots,X_n)$. Then by Theorem 1 there is an ordering $\leq$ of K which makes f negative. Therefore $\leq$ extends to the field

$L = K(\sqrt{-f})$. Since L is the function field of the variety V defined by $1 + Y^2 \cdot f(X_1,\dots,X_n) = 0$, we find some $a_1,\dots,a_n \in \mathbb{R}$ with $f(a_1,\dots,a_n) < 0$ by Theorem 2. Thus f is not positive definite.

In the course of his solution, Artin together with Schreier [A-S] developed the general theory of formally real fields. A field K is called formally real if it admits at least one ordering.

Using Zorn's Lemma, it becomes clear that every ordered field $(K,\leq)$ admits a maximal algebraic extension field to which the ordering $\leq$ also extends. Such an extension is usually called a real closure of $(K,\leq)$. In case $(K,\leq)$ does not have any proper algebraic order-extension field, it is called real closed. The main theorems concerning algebraic order-extensions are the Characterization Theorem for real closed fields

THEOREM 3 *An ordered field* $(K,\leq)$ *is real closed if and only if (i) every odd degree polynomial* $f \in K[X]$ *has a zero in* K, *and (ii) every element or its negative is a square in* K,

and the Isomorphism Theorem for real closures

THEOREM 4 *Any two real closures of an ordered field* $(K,\leq)$ *are isomorphic over* K.

Theorem 2 remains valid if we replace $\mathbb{R}$ by any real closed field R. With this generalization of Theorem 2 in mind, we may call Theorems 1-4 the basic theorems of the theory of formally real fields.

Before we proceed to the 'p-adic' case let us briefly sketch the model theory of real closed fields (assuming familiarity with model theoretic notions). From the above mentioned generalization of Theorem 2, using Robinson's Test, one deduces the model completeness of the theory of real closed fields, i.e.

THEOREM 5 Let $(K,\leq)$ and $(L,\leq)$ be real closed and $K \subset L$. Then an elementary sentence about ordered fields with parameters from K holds in $(L,\leq)$ if and only if it holds in $(K,\leq)$.

Using this theorem and the fact that the algebraic closure of $\mathbb{Q}$ in an arbitrary real closed field is again real closed one obtains Tarski's Transfer Principle: every elementary sentence about ordered fields which holds in $(\mathbb{R},\leq)$ also holds in any real closed field $(K,\leq)$. This principle is equivalent to

THEOREM 6 Every elementary sentence about ordered fields which holds in $(\mathbb{R},\leq)$ is a logical consequence of the axioms of real closed fields (i.e. field axioms + $o_1 - o_6$ + (i) and (ii)).

In particular, this theorem gives an effective method in deciding whether an elementary sentence about ordered fields holds in $(\mathbb{R},\leq)$ or not.

It was this decidability which motivated Ax-Kochen [A-K] and Eršov [Er] to try the same for the p-adic number field $\mathbb{Q}_p$. They succeeded and, in the course of their proof for the decidability, they also introduced the notion of a p-adically closed field. Only later, Kochen [K] developed the theory of p-valued fields in analogy to ordered fields. He was also searching for an analog of Hilbert's 17th Problem. This search was completed by the second author. In our presentation here, we are trying to follow the development of the real case more closely.

To begin with, let us look first for an analog of the notion 'ordering' in the p-adic case. The above defined integer $v_p(x)$ yields a valuation $v_p : \mathbb{Q} \to \mathbb{Z} \cup \{\infty\}$ (where we let $v_p(0) = \infty$) on the rational number field $\mathbb{Q}$ which extends canonically to its completion $\mathbb{Q}_p$. The valuation ring attached to v_p is

$$\mathcal{O}_p = \{x \mid v_p(x) \geq 0\}.$$

We may also define the following divisibility relation on $\mathbb{Q}_p$

$$x \mid_p y \quad \text{iff} \quad ax = y \quad \text{for some} \quad a \in \mathcal{O}_p .$$

Clearly, $x \mid_p y$ is equivalent to $v_p(x) \leq v_p(y)$ and an element $x \in K$ belongs to $\mathcal{O}_p$ if and only if $1 \mid_p x$. In listing some basic rules for $\mid_p$ we omit the index p at the same time.

d_1 : $x \mid x$

d_2 : $x \mid y \, , \, y \mid z \Rightarrow x \mid z$

d_3 : $x \mid y$ or $y \mid x$

d_4 : $x \mid y \Rightarrow xz \mid yz$

d_5 : $1 \mid x \, , \, 1 \mid y \Rightarrow 1 \mid x+y$

d_6 : $p \nmid 1$

d_7 : $1 \mid x \Rightarrow p \mid (x-1)$ or ... or $p \mid (x-p)$

We now call a binary relation $\mid$ on a field K of characteristic zero a <u>p-divisibility</u> if it satisfies d_1-d_7 . If $\mathcal{O}$ is the set of $x \in K$ with $1 \mid x$, it is easy to see that d_1-d_5 imply $\mathcal{O}$ to be a valuation ring. Let v be a valuation of K with $\mathcal{O}$ as its valuation ring, then d_6 and d_7 imply

(1) $v(p)$ is minimal positive in the value group

(2) the residue class field is $\mathbb{Z}/p\mathbb{Z}$.

A valuation v satisfying (1) and (2) will be called a p-valuation. (Later this definition will be generalized to p-valuations of certain p-rank d; what we now call a p-valuation will have p-rank 1 later.) It is easily shown that, if we start with a p-valuation v on K and define $|$ in the obvious way, then $|$ satisfies $d_1 - d_7$. Therefore we may sometimes replace a p-divisibility by a p-valuation and conversely. A field K of characteristic 0 together with a p-valuation v will be called a p-valued field.

As in the real case it is now possible to prove, using the theorems below, that any other 'rule' of the relation $|_p$ on $\mathbb{Q}_p$ is a logical consequence of $d_1 - d_7$. (The proof will be given at the end of this section.) Thus we may consider a p-divisibility (or equivalently a p-valuation) as the counterpart of an ordering in the p-adic case. We may therefore expect similar theorems to hold in the theory of formally p-adic fields. A field K of characteristic 0 is called formally p-adic if it admits at least one p-valuation.

In the analog of Theorem 1 we are looking for the set of elements in a field K which are integral for each p-valuation of K, i.e. we have to describe the intersection of all p-valuation rings. To find such a description we look first for an operator γ which does in the p-adic case what the square-operator did in the real case. In the real case X^2 is positive for every ordering of K . In the p-adic case

$$\gamma(X) = \frac{1}{p} \cdot \frac{X^p - X}{(X^p - X)^2 - 1}$$

is integral for every p-valuation of K . This operator is called the <u>Kochen operator</u>. Now the analog of Theorem 1 is

THEOREM 1_p <u>The intersection of all p-valuation rings of</u> K <u>is equal to the ring of quotients of the form</u> $\frac{b}{1+pc}$ <u>where</u> $b,c \in \mathbb{Z}[\gamma K]$ = <u>ring generated by the elements</u> $\gamma(x)$ <u>with</u> $x \in K$. (Theorem 6.14)

Concerning Theorem 2 we have the following analog:

THEOREM 2_p <u>Let</u> V <u>be an affine variety defined over</u> $\mathbb{Q}_p$. <u>Then</u> V <u>admits a non-singular p-adic point if and only if the function field</u> $\mathbb{Q}_p(V)$ <u>admits a p-valuation</u> (cf. Theorem 7.8).

Using Zorn's Lemma it is again clear that every p-valued field (K,v) admits a maximal algebraic p-valued extension. Such an extension will be called a <u>p-adic closure</u> of (K,v). If a p-valued field (K,v) does not have any proper algebraic p-valued extension, it is called <u>p-adically closed</u>.

In the analog of Theorem 3 the counterpart of (i) will be <u>Hensel's Lemma</u> which, in its simplest form (compare [Ri], p.185) for a p-valued field (K,v) says that every polynomial of the type

$$X^{n+1} + X^n + p\, a_{n-1}\, X^{n-1} + \dots + p\, a_o$$

with $v(a_i) \geq 0$ has a zero in K. Property (ii) of Theorem 3 expresses that for all $x \in K$ either $+x$ or $-x$ is a square. In the p-adic case one has to consider all n-th powers rather than just squares. Again it is possible to find a set of representatives $\rho_1^{(n)}, \dots, \rho_{m_n}^{(n)}$ in $\mathbb{N}$ such that in a p-adically closed field (K,v) for each $x \in K$ one of the elements $\rho_i^{(n)} \cdot x$ (with $1 \leq i \leq m_n$) is an n-th power. The property of (K,v) responsible for this is just a property of the value group $\Gamma = vK$, namely: <u>for every</u> $n \in \mathbb{N}$ <u>and every</u> $\gamma \in \Gamma$ <u>one of the elements</u> $\gamma - 1, \dots, \gamma - n$ <u>is divisible by</u> n <u>in</u> Γ.

An ordered group Γ which has a minimal positive element 1 satisfying this property is called a <u>$\mathbb{Z}$-group</u> (since the elementary theory of the ordered group $\mathbb{Z}$ is determined by this property). Now the analog of the <u>Characterization Theorem</u> is

THEOREM 3_p <u>A p-valued field</u> (K,v) <u>is p-adically closed if and only if</u> $(i)_p$ <u>Hensel's Lemma holds for</u> (K,v), <u>and</u> $(ii)_p$ <u>the value group</u> vK <u>is a $\mathbb{Z}$-group</u> (cf. Theorem 3.1).

The p-adic analog of the <u>Isomorphism Theorem</u> is

THEOREM 4_p <u>Let</u> (K,v) <u>be p-valued. Two p-adic closures</u> (L_1,v_1) <u>and</u> (L_2,v_2) <u>of</u> (K,v) <u>are isomorphic over</u> K <u>if and only if</u> $L_1^n \cap K = L_2^n \cap K$ <u>for all</u> $n \geq 2$ <u>where</u> L_i^n <u>is the set of</u> n-<u>th powers of elements of</u> L_i (cf. Corollary 3.11).

Although in the p-adic case in general there is not just one p-adic closure, we may still call this theorem an analog: In the real case the condition $L_1^n \cap K = L_2^n \cap K$ is always satisfied, since

$$L_1^{2n+1} \cap K = K = L_2^{2n+1} \cap K$$

and

$$L_1^{2n} \cap K = \{x \in K \mid 0 \leq x\} = L_2^{2n} \cap K .$$

Thus in the real case the information provided by the ordering of K already determines the real closure, while in the p-adic case the information provided by the p-valuation in general is not sufficient to determine the closure.

As in the real case Theorem 2_p not just holds for $\mathbb{Q}_p$ but for every p-adically closed field instead of $\mathbb{Q}_p$. This implies that the complete analogs of Theorem 5 and 6 hold in the p-adic case:

THEOREM 5_p Let (K,v) and (L,v) be p-adically closed and $K \subset L$. Then an elementary sentence about valued fields with parameters from K holds in (L,v) if and only if it holds in (K,v) (cf. Theorem 5.1).

THEOREM 6_p Every elementary sentence about valued fields which holds in $(\mathbb{Q}_p,v_p)$ is a logical consequence of the axioms of p-adically closed fields (i.e. field axioms + char = 0 + d_1-d_7 + $(i)_p$ and $(ii)_p$).

In particular, we also have a transfer principle from $(\mathbb{Q}_p,v_p)$ to every p-adically closed field (L,v) for elementary sentences about valued fields. Therefore every 'rule' for $|_p$ valid in $\mathbb{Q}_p$ transfers to L with respect to the p-divisibility $|_v$ defined by v. This is obvious if we define a rule to be a sentence of the type

$$\forall x_1,\ldots,x_m \; [\alpha(x_1,\ldots,x_m) \Rightarrow \beta(x_1,\ldots,x_m)]$$

where the quantification $\forall x_1,\ldots,x_m$ is to be taken over all field elements, and α and β are boolean combinations (using "and" and "or") of atomic expressions like $f(x_1,\ldots,x_m) = 0$ and $g(x_1,\ldots,x_m) \mid h(x_1,\ldots,x_m)$ with $f,g,h \in \mathbb{Z}[x_1,\ldots,x_m]$. From the shape of a rule we see that it holds in a p-valued field (K,v) whenever it holds in some p-adic closure (L,v) of (K,v). Thus a rule valid for $|_p$ in $\mathbb{Q}_p$ transfers to all p-valued fields, hence it is a logical consequence of the axioms for such fields. So we finally see that our choice of axioms for a p-divisibility (and hence for a p-valuation) is the right one.

§ 2. p-valuations

2.1 Definitions and examples

As already mentioned in the introduction, we will not only study the algebraic theory of the fields $\mathbb{Q}_p$ but also that of their extension fields K of finite degree $d = [K:\mathbb{Q}_p]$. Accordingly, the notion of p-valued fields has to be defined in such a generality as to cover those fields. Before doing so, let us fix some notations which will be used throughout this book. If

K is a valued field (note that we no longer mention the valuation explicitly), then

v will usually denote its valuation, hence

$v(a)$ is the value of an element $a \in K$. We often will omit the brackets and write va in order to simplify notation.

vK denotes the value group, consisting of all values va with $0 \neq a \in K$;

$\mathcal{O}$ is the valuation ring, consisting of all $a \in K$ with $va \geq 0$;

$\mathfrak{M}$ is the maximal ideal of $\mathcal{O}$;

$\bar{K}$ $= \mathcal{O}/\mathfrak{M}$ the residue field;

$\bar{a}$ the residue class in $\bar{K}$ of an element $a \in \mathcal{O}$.

By definition the value group vK is a totally ordered abelian group, written additively, and v is a homomorphism of the multiplicative group $K^{\times}$ onto vK such that

$$va \geq vb \iff ab^{-1} \in \mathcal{O} .$$

The kernel of this homomorphism is the unit group $\mathcal{O}^{\times}$ of $\mathcal{O}$; hence we may identify $vK = K^{\times}/\mathcal{O}^{\times}$ if convenient. After such an identification va becomes the residue class of a modulo $\mathcal{O}^{\times}$.

Now let p be a prime number. We suppose the residue field $\bar{K}$ to be of characteristic p while K <u>should not be of characteristic</u> p , hence $\mathrm{char}(K) = 0$. Given any natural number d , the condition

$$\dim \mathcal{O}/p = d$$

defines the <u>p-valued fields of p-rank d</u> . The dimension is to be understood over the field $\mathbb{Z}/p\mathbb{Z}$ of p elements.

The above condition implies that the vector space $\mathcal{O}/p$ is finite and therefore has only finitely many subspaces. In particular there are only finitely many ideals in $\mathcal{O}$ which contain p . Now the principal $\mathcal{O}$-ideals $\mathcal{O}a$ containing p correspond 1-1 to the values $va \in vK$ such that $0 \leq va \leq vp$. We conclude that there are only finitely many positive elements in vK which are $\leq vp$. The number e of these elements is called the p-<u>ramification index</u> of K . Among the e positive elements $\leq vp$ consider the smallest one, say $v\pi$ with $\pi \in \mathfrak{M}$. By definition $v\pi$ is the smallest positive element in vK . The maximal ideal $\mathfrak{M}$ is generated by π . We say that π is a <u>prime element</u> of the p-valued field K .

Consider the cyclic subgroup of vK generated by $v\pi$; this subgroup will be denoted by $\mathbb{Z}\cdot v\pi$. We often identify

$$v\pi = 1 \ ;$$

then $\mathbb{Z}\cdot v\pi = \mathbb{Z}$ and $\mathbb{Z}$ becomes an ordered subgroup of vK, the element $1 \in \mathbb{Z}$ being the smallest positive element in vK. The latter property implies that $\mathbb{Z}$ is a convex subgroup of vK. This means that if $\alpha \in vK$ is bounded by integers, i.e. if $m \leq \alpha \leq n$ with $m,n \in \mathbb{Z}$ then $\alpha \in \mathbb{Z}$.

It follows that the e positive elements $\leq vp$ in vK are precisely the first e positive integers $1,2,\ldots,e$ and that

$$vp = e \quad .$$

In particular we see that $vp \in \mathbb{Z}$ and that $\mathbb{Z}$ is the smallest convex subgroup of vK containing vp.

Sometimes, if considering several fields which may have different p-ramification indices, it will be convenient to use a universal normalization, independent of the field. In those cases we identify

$$vp = 1$$

and hence

$$v\pi = \frac{1}{e} \quad .$$

With this kind of normalization the group $\mathbb{Z}\cdot\frac{1}{e}$ is convex in vK while $\mathbb{Z}$ itself is not convex (except if $e = 1$).

Now consider the descending sequence of $\mathcal{O}$-ideals:

$$\mathcal{O} \supset \mathcal{O}\pi \supset \mathcal{O}\pi^2 \supset \ldots \supset \mathcal{O}\pi^e = \mathcal{O}p.$$

Each factor $\mathcal{O}\pi^{i-1}/\mathcal{O}\pi^i$ is naturally isomorphic (as vector space) to $\mathcal{O}/\mathcal{O}\pi = \mathcal{O}/\mathfrak{M} = \bar{K}$, the isomorphism being given by multiplication with π^{i-1} . Hence

$$d = \dim \mathcal{O}/p = e \cdot \dim \bar{K} = e \cdot f$$

where $f = \dim \bar{K}$ is the <u>residue degree</u> of K , i.e. the degree of $\bar{K}$ over its prime field. The formula

$$d = e \cdot f$$

puts into evidence that in a p-valued field, the p-ramification index e and the residue degree f are both finite. Conversely, if K is any valued field of characteristic zero and residue characteristic p , and if the p-ramification index e and residue degree f of K are finite, then $d = \dim \mathcal{O}/p$ is finite too and hence K is p-valued according to our definition above.

We are now going to give a series of examples for p-valued fields. Clearly, every finite extension of any of the examples leads to others (of possibly higher p-rank) just by considering all the extensions of the given p-valuation on the ground field.

<u>Example 2.1</u>: The most canonical example is, of course, the p-valuation v_p on the p-adic number field $\mathbb{Q}_p$. At the same time the restriction of v_p to any subfield of $\mathbb{Q}_p$ (in particular $\mathbb{Q}$) is an example too. Clearly, all these p-valuations have p-rank 1.

<u>Example 2.2</u>: Let v be any p-valuation of p-rank d on a field K; e.g. v_p on $\mathbb{Q}_p$. Consider the rational function field $F = K(X)$ in one variable. We will define a p-valuation w on F with values in the group $\Gamma = \mathbb{Z} \times vK$ <u>ordered lexicographically</u>, i.e. for all (a,b) and $(a',b') \in \Gamma$ we have

$$(a,b) < (a',b') \quad \text{iff} \quad a < a' \quad \text{or } (a=a' \text{ and } b < b') .$$

The element $(0,v\pi)$ is minimal positive and $\mathbb{Z}\cdot v\pi$ is a convex subgroup of Γ. For the polynomial

$$f(X) = a_n X^n + a_{n+1} X^{n+1} + \ldots + a_m X^m \in K[X]$$

with $a_n \neq 0$ and $n \leq m$ we define

$$w(f) = (n, va_n) .$$

One easily checks that w yields a p-valuation on F of the same p-rank as v. Clearly w extends v.

<u>Example 2.3</u>: We extend the definition of the last example to <u>formal Laurent series</u> over K. For the formal series

$$f(X) = a_n X^n + a_{n+1} X^{n+1} + \ldots = \sum_{i=n}^{\infty} a_i X^i$$

we define again $w(f) = (n, va_n)$, if $a_n \neq 0$. But now n is any integer, positive or negative. It is again easy to check that w is a p-valuation on the field $K((X))$ of formal Laurent series. Clearly w extends v and is of the same p-rank.

<u>Example 2.4</u>: We extend the definition of w further to <u>Puiseux series</u> over K, i.e. to series

$$f(X) = \sum_{i=n}^{\infty} a_i X^{\frac{i}{m}} \qquad \text{with } n \in \mathbb{Z} \text{ and } m \in \mathbb{N} .$$

If $a_n \neq 0$ we let $w(f) = (\frac{n}{m}, va_n)$. Now the group $\mathbb{Q} \times vK$, ordered lexicographically, acts as group of values. By this we obtain a p-valuation w of p-rank d on the field $\bigcup_{m \in \mathbb{N}} K((X^{\frac{1}{m}}))$ of Puiseux series over K.

Example 2.5: Let v be a p-valuation of p-rank d on K. Let $\mathcal{O}_v = \{x \mid v(x) \geq 0\}$ be its valuation ring. Assume $\varphi: F \to K \cup \{\infty\}$ is some place of F onto K, e.g. let φ be a $\mathbb{Q}_p$-place (i.e. a place into $\mathbb{Q}_p$), K its image, and v the restriction of v_p to K. Then it is easy to check that $\varphi^{-1}(\mathcal{O}_v)$ is a valuation ring on F determining a p-valuation of p-rank d. If we compare this example with Example 2.2, we find that the valuation ring of w in 2.2 is equal to the preimage of $\mathcal{O}_v$ under the place determined by $X \mapsto 0$ on $K(X)$.

Example 2.6: For the reader familiar with model theory we could just say: take any non-trivial ultra-power of a p-valued field K of p-rank d in order to obtain a proper extension of the same p-rank. For the reader not familiar with ultra-powers we will introduce them here, restricted to valued fields. We choose an approach which avoids the use of formulas. This example will be used later in Section 7 for the proof of the existence of rational places (see Theorem 7.2 and Proposition 7.3)

Let K be any field (of characteristic 0 as we agreed at the beginning) and $\mathcal{O}$ a valuation ring of K. Defining addition, subtraction and multiplication componentwise for sequences

$$a = (a^{(n)})_{n \in \mathbb{N}} = (a^{(1)}, a^{(2)}, \ldots, a^{(n)}, \ldots)$$

with $a^{(n)} \in K$, the direct product $\prod_{n \in \mathbb{N}} K$ becomes a ring. Clearly, the direct sum

$$\sum_{n \in \mathbb{N}} K = \{a \in \prod_{n \in \mathbb{N}} K \mid a^{(n)} \neq 0 \text{ for only finitely many } n\}$$

is an ideal in ΠK . Let $M \supset \Sigma K$ be any maximal ideal of ΠK . Then the quotient

$$K^* = \Pi K/M$$

is a field. For a sequence $a \in \Pi K$ let us call

$$Z(a) = \{n \mid a^{(n)} = 0\}$$

the <u>zero-support</u> of a . Forming the set

$$D = \{Z(a) \mid a \in I\}$$

for an arbitrary proper ideal I of the ring ΠK one easily checks that it constitutes a filter on $\mathbb{N}$. In case I is a maximal ideal, D is an <u>ultra-filter</u>. If ΣK is contained in I , every cofinite subset of $\mathbb{N}$ belongs to D . Thus, for our maximal ideal M , chosen above, we know that D is an ultra-filter on $\mathbb{N}$ which is <u>non-principal</u>, i.e. does not contain any finite subset of $\mathbb{N}$. Moreover, one checks that for sequences $a,b \in \Pi K$

$$a \equiv b \mod M \quad \text{iff} \quad \{n \mid a^{(n)} = b^{(n)}\} \in D .$$

We now introduce a valuation ring $\mathcal{O}^*$ on K^* by defining

$$a + M \in \mathcal{O}^* \quad \text{iff} \quad \{n \mid a^{(n)} \in \mathcal{O}\} \in D$$

for $a \in \Pi K$. This definition is independent of the choice of the representative a . Indeed, let $a \equiv b \mod M$ then

$$\{n \mid a^{(n)} = b^{(n)}\} \cap \{n \mid a^{(n)} \in \mathcal{O}\} \in D$$

and, since $\{n \mid b^{(n)} \in \mathcal{O}\}$ contains this intersection, also

$\{n|b^{(n)} \in \mathcal{O}\} \in D$. We leave it as an exercise to the reader to check that $\mathcal{O}^*$ defines a p-valuation of p-rank d on K^*. As an example for such a proof let us show that $\mathcal{O}^*$ satisfies

$$ab = 1 \quad \Rightarrow \quad a \in \mathcal{O}^* \quad \text{or} \quad b \in \mathcal{O}^*.$$

More precisely, by $ab = 1$ we mean $ab \equiv 1 \bmod M$ with $a,b \in \Pi K$. Hence $\{n|a^{(n)} \cdot b^{(n)} = 1\} \in D$. Since $\mathcal{O}$ is a valuation ring of K we know that

$$\{n|a^{(n)} \cdot b^{(n)} = 1\} \subset \{n|a^{(n)} \in \mathcal{O}\} \cup \{n|b^{(n)} \in \mathcal{O}\} .$$

From the ultra-filter properties of D it follows that

$$\{n|a^{(n)} \in \mathcal{O}\} \in D \quad \text{or} \quad \{n|b^{(n)} \in \mathcal{O}\} \in D .$$

Hence $a + M \in \mathcal{O}^*$ or $b + M \in \mathcal{O}^*$, as we have claimed.

We have not used so far that D is non-principal. This will be important only later. Note that K embeds into K^* by

$$\alpha \to (\alpha, \alpha, \ldots) \in \Pi K$$

for $\alpha \in K$. Clearly, the valuation ring $\mathcal{O}^*/M$ of K^* extends the given valuation ring $\mathcal{O}$ of K .

2.2 Some valuation theory

We will need later some facts from general valuation theory, primarily about Henselian fields and decomposition of valuations. For the convenience of the reader we will discuss those facts in this subsection. As a general reference for valuation theory let us mention [Ax], [E], and [Ri].

Let us discuss first the Henselian property. A valued field K is called Henselian if every algebraic field extension admits only one valuation extending the given valuation of K. It is well-known that this is equivalent to the validity of

HENSEL'S LEMMA Let $f(X) \in \mathcal{O}[X]$ and suppose that there is some $a \in \mathcal{O}$ such that the residue class of a is a simple zero of the image of f in the residue class field. Then there is $t \in K$ with $f(t) = 0$ and $v(t-a) > 0$.

It suffices to check this for polynomials of the type (cf. [Ri], p.186)

$$f(X) = X^{n+1} + X^n + b_{n-1}X^{n-1} + \ldots + b_o$$

with $b_o, \ldots, b_{n-1} \in \mathfrak{M}$. There are still other equivalent formulations, like Krasner's Lemma or the Hensel-Rychlik Lemma or the Newton Lemma. Here we prefer Newton's Lemma since we have to use it in Section 3. The statement of the lemma refers to a valued field K with valuation v and valuation ring $\mathcal{O}$, and it reads as follows:

NEWTON'S LEMMA Let $f(X) \in \mathcal{O}[X]$ and denote its derivative by $f'(X)$. Suppose there exists $a \in \mathcal{O}$ such that $vf(a) > 2 \cdot vf'(a)$. Then $f(X)$ admits a root in K. In fact, there is one and only one $t \in K$ such that $f(t) = 0$ and $v(t-a) > vf'(a)$.

Hensel's Lemma in the original sense is contained here as a special case, namely if $vf'(a) = 0$.

According to general valuation theory any valued field K admits a unique smallest Henselian valued extension field; this is called the <u>Henselization</u> of K and denoted by K^h. By definition K^h is a Henselian valued field extension of K with the following universal mapping property: given any Henselian valued field extension L of K then there exists a unique K-isomorphic embedding of K^h into L (as valued fields). Usually we shall identify $K^h \subset L$.

The Henselization K^h has the following property: K^h is separably algebraic over K. As a valued field, K^h is an <u>immediate</u> extension of K; this means that K^h and K have the same value group and the same residue field:

$$vK^h = vK \quad \text{and} \quad \overline{K^h} = \bar{K} \ .$$

If K is p-valued of p-rank d then it follows that K^h is p-valued too, of the same p-rank d. A prime element $\pi \in K$ remains a prime element in K^h. If $u_1, \ldots, u_d \in K$ form a $\mathbb{Z}/p$ - basis of $\mathcal{O}$ modulo p then they retain this property in K^h with respect to the valuation ring $\mathcal{O}^h$ of K^h.

We shall have to use the above properties of K^h in our discussion in the next sections.

It is well-known that the completion $K_{\mathfrak{P}}$ of a number field K of finite degree over $\mathbb{Q}$ w.r.t. a finite prime $\mathfrak{P}$ is a Henselian field. It contains (an isomorphic copy of) the completion $\mathbb{Q}_p$ of the rationals $\mathbb{Q}$ with respect to the prime number p induced by $\mathfrak{P}$. The canonical p-valuation of $K_{\mathfrak{P}}$ has p-rank $d = [K_{\mathfrak{P}} : \mathbb{Q}_p]$.

Now let us consider <u>extensions of valued fields</u>. If $L|K$ is an extension of valued fields then usually we use the same symbol v to denote the valuation on K and that on L. (This is in analogy to the notation used in the theory of ordered fields where the same symbol $\leq$ is used to denote the order relation of a field and of its ordered extension fields.) If we want to consider other objects connected with the valuation then we use subscripts to indicate which field we are considering. Accordingly we write $\mathcal{O}_K$ resp. $\mathcal{O}_L$ for the valuation rings. We have

$$K \cap \mathcal{O}_L = \mathcal{O}_K .$$

We always use the natural identification

$$vK \subset vL \quad \text{and} \quad \bar{K} \subset \bar{L}$$

and

$$\mathcal{O}_K/p \subset \mathcal{O}_L/p \quad .$$

Suppose K is p-valued of p-rank $d_K = \dim \mathcal{O}_K/p$. If $[L:K]$ is finite then we claim that L is p-valued and

$$d_K \leq d_L \quad .$$

In order to prove this and also to compute d_L we proceed as follows.

The p-rank $d_L = \dim \mathcal{O}_L/p$ is finite if and only if the p-ramification index e_L and the residue degree f_L are both finite. As to the latter we have by definition

$$f_L = f_K \cdot [\bar{L}:\bar{K}] \ .$$

If $[L:K]$ is finite then it is well known from general valuation theory that

$$[\bar{L}:\bar{K}] \leq [L:K] \quad ;$$

it follows that f_L is finite .

Let $[vL:vK]$ denote the index of the value groups. If $[L:K]$ is finite then we know from valuation theory that even the following inequality holds:

$$[vL:vK]\cdot[\bar{L}:\bar{K}] \leq [L:K] \;;$$

it follows that $[vL:vK]$ is finite. We also have to consider the <u>initial index</u> $i[vL:vK]$, defined to be the number of positive elements $\beta \in vL$ which are $\leq$ all positive elements $\alpha \in vK$. This means that $\beta \leq v\pi$ where π is a prime element of K . If β,β' are two different such elements then they are contained in different cosets modulo vK . For suppose that $0<\beta<\beta' \leq v\pi$, then $0<\beta'-\beta < \beta' \leq v\pi$ hence $\beta'-\beta \notin vK$ since $v\pi$ is the smallest positive element in vK . We conclude that the initial index is not larger than the relative ramification index:

$$i[vL:vK] \leq [vL:vK] \quad .$$

In particular, if $[L:K]$ is finite then $i[vL:vK]$ is finite too. We claim that

$$e_L = e_K\cdot i[vL:vK] \; .$$

Let us put $i = i[vL:vK]$. Among the i positive elements $\beta \in vL$ with $\beta \leq v\pi$ consider the smallest one, say $v\Pi$ with $\Pi \in L$. Then $v\Pi$ is the smallest positive element in the whole group vL, i.e. Π is a prime element in L . Let us identify $v\Pi = 1$; then

$\mathbb{Z}$ becomes a convex subgroup of vL. The i positive elements $\beta \leq v\pi$ are now precisely the i integers $1,2,\ldots,i$ and we have

$$v\pi = i .$$

Hence

$$vp = e_K \cdot v\pi = e_K \cdot i$$

which shows that

$$e_L = e_K \cdot i$$

as contended. For later reference, let us formulate the result of this discussion in the following

LEMMA 2.7 _Let_ K _be a p-valued field and_ L _a valued field extension of_ K. _The necessary and sufficient condition for_ L _to be p-valued is that the initial index_ $i[vL:vK]$ _and the residue degree_ $[\bar{L}:\bar{K}]$ _are both finite. If this is so then_

$$e_L = e_K \cdot i[vL:vK] \quad , \quad f_L = f_K \cdot [\bar{L}:\bar{K}] .$$

Hence the p-rank d_L _is given by_

$$d_L = d_K \cdot i[vL:vK] \cdot [\bar{L}:\bar{K}] .$$

_If_ $[L:K]$ _is finite then the above condition is always satisfied; we have_

$$i[vL:vK] \cdot [\bar{L}:\bar{K}] \leq [vL:vK] \cdot [\bar{L}:\bar{K}] \leq [L:K] .$$

The initial index $i[vL:vK]$ is actually a divisor of $[vL:vK]$. To put this into evidence and also to interpret its complementary divisor, we consider the _canonical decomposition_ of a p-valuation, as follows.

For a moment we work in K only and hence we identify $v\pi = 1$, so that $\mathbb{Z}$ becomes a convex subgroup of vK. Let us put

$$\dot{v}K = vK/\mathbb{Z} \quad ;$$

if $0 \neq a \in K$ then $\dot{v}a$ should denote the coset of va modulo $\mathbb{Z}$. Since $\mathbb{Z}$ is convex in vK, the factor group $vK/\mathbb{Z}$ inherits from vK the structure of a totally ordered group. Therefore the map

$$\dot{v} : K^{\times} \to \dot{v}K$$

is a valuation of the field K; it is called the <u>coarse valuation</u> belonging to v, the coarsening referring to the convex subgroup $\mathbb{Z} \subset vK$ which is characterized as the smallest convex subgroup of vK containing vp. The valuation ring $\dot{\mathcal{O}}$ of $\dot{v}$ in K is characterized as the smallest overring of $\mathcal{O}$ in which p becomes a unit; hence $\dot{\mathcal{O}}$ is the ring of quotients of $\mathcal{O}$ with respect to powers of p as admissible denominators:

$$\dot{\mathcal{O}} = \mathrm{Quot}_p(\mathcal{O}) = \{\frac{a}{p^n} : a \in \mathcal{O} , n \in \mathbb{N}\} .$$

It may well be that $\dot{v}$ is the trivial valuation; this is the case if and only if $vK = \mathbb{Z}$.

Let $\dot{\mathfrak{M}}$ denote the maximal ideal of $\dot{\mathcal{O}}$; then

$$\dot{\mathfrak{M}} \subset \mathfrak{M} \subset \mathcal{O} \subset \dot{\mathcal{O}} .$$

We have to introduce a new symbol for the residue field $\dot{\mathcal{O}}/\dot{\mathfrak{M}}$ (we cannot write $\bar{K}$ since this denotes the residue field $\mathcal{O}/\mathfrak{M}$ of K with respect to v). Let us put

$$K^{o} = \dot{\mathcal{O}}/\dot{\mathfrak{M}} \quad .$$

For $a \in \dot{\mathcal{O}}$ let a° be its residue in K° . The field K° is called the <u>core field</u> of the p-valuation v of K . The core field carries naturally a valuation whose valuation ring is the image of $\mathcal{O}$:

$$\mathcal{O}^{\circ} = \mathcal{O}/\dot{\mathfrak{m}} \quad .$$

The valuation of K° belonging to this valuation ring can be explicitly given as follows. Let $0 \neq a^{\circ} \in K^{\circ}$. Let $a \in \dot{\mathcal{O}}$ be a foreimage of a°. Since $a^{\circ} \neq 0$ we have $\dot{v}a = 0$, hence $va \in \mathbb{Z}$. Now it is a straightforward verification that the value $va \in \mathbb{Z}$ depends on a° only, and not on the choice of its foreimage a . Moreover the map $a^{\circ} \mapsto va$ is a valuation of K° with value group $\mathbb{Z}$, whose valuation ring is precisely $\mathcal{O}^{\circ}$. Again we use the symbol v to denote the valuation on K° ; thus we write

$$va^{\circ} = va \qquad\qquad (a^{\circ} \in K^{\circ})$$

if a is a foreimage of a° . With this notation the value group of K° is denoted by vK° , and we have seen that

$$vK^{\circ} = \mathbb{Z} \quad ,$$

provided we identify $\mathbb{Z}$ with the convex subgroup of vK as explained above. If this identification is not made then

$$vK^{\circ} = \mathbb{Z}\cdot v\pi$$

and vK° can be characterized as the smallest convex subgroup of vK containing vp .

By definition of the valuation ring $\mathcal{O}^{\circ}$ we have trivially:

$$\overline{K^{\circ}} = \mathcal{O}^{\circ}/\mathfrak{m}^{\circ} = \mathcal{O}/\mathfrak{m} = \bar{K} \quad .$$

That is, the core field K^o has the same residue field as K .

We conclude that the core field K^o is p-valued, with the same p-ramification index and residue degree as K; also char $K^o = 0$. Hence the p-ranks of K and K^o also coincide: $d_K = d_{K^o}$.

Our discussion has shown:

Let K be a p-valued field. The given p-valuation v of K is canonically decomposed into its coarse valuation and its core valuation. The coarse valuation $\dot{v}$ is again a valuation of K ; its valuation ring equals the ring of quotients of $\mathcal{O}$ with respect to the powers of p as admissible denominators. Let K^o denote the residue field of K with respect to the coarse valuation. K^o is called the core field of K . The core valuation is a valuation of K^o ; it is again denoted by v . The value group vK^o is a convex subgroup of vK , namely the smallest convex subgroup containing vp . In particular $vK^o \simeq \mathbb{Z}$. The value group of the coarse valuation $\dot{v}$ can now be described as

$$\dot{v}K = vK/vK^o \quad .$$

The core field K^o is p-valued; it has the same p-ramification index as K , the same residue degree and hence the same p-rank.

This being said, consider again a valued field extension L of K . We suppose that L is p-valued; if $[L:K]$ is finite this is always the case by Lemma 2.7 . Due to the canonical construction of the coarse valuation it is clear that the coarse valuation of L is a prolongation of the coarse valuation of K ;

hence both will be denoted by the same symbol $\dot{v}$. The core field L^o is an extension of the core field K^o , not only as a field but also as a valued field. Consider the relative ramification index $[vL^o:vK^o]$ of the core field extension. We claim that this equals the initial index $i[vL:vK]$. Indeed: we know that vL^o is the smallest convex subgroup of vL containing vp ; hence $vL^o = \mathbb{Z}\cdot v\Pi$ where Π is a prime element of L . Similarly $vK^o = \mathbb{Z}\cdot v\pi$. Since $v\pi = i\cdot v\Pi$ (with $i = i[vL:vK]$) we conclude

$$[vL^o:vK^o] = [\mathbb{Z}:i\mathbb{Z}] = i \quad .$$

Note that

$$[vL:vK] = [\dot{v}L:\dot{v}K]\cdot[vL^o:vK^o] \ .$$

This is so because

$$\dot{v}L = vL/vL^o \quad \text{and} \quad \dot{v}K = vK/vK^o$$

and by definition:

$$vL^o \cap vK = vK^o \ .$$

We obtain:

LEMMA 2.8 *Let* $L|K$ *be an extension of* p*-valued fields. Then the initial index* $i[vL:vK]$ *equals the relative ramification index of the core fields*:

$$i[vL:vK] = [vL^o:vK^o] \ .$$

Also we have the formula:

$$[vL:vK] = [\dot{v}L:\dot{v}K]\cdot[vL^o:vK^o]$$

showing that the initial index is a divisor of $[vL:vK]$, *the complementary divisor being* $[\dot{v}L:\dot{v}K]$.

For simplicity, the following lemma will be restricted to the case of Henselian base fields (otherwise we would have to consider not only one prolongation, but all prolongations of the valuation of K to the finite extension field L) .

LEMMA 2.9 Let $L|K$ be a finite extension of p-valued fields. Suppose that K is Henselian. Then the fundamental degree formula holds:

$$[L:K] = [vL:vK]\cdot[\bar{L}:\bar{K}] .$$

In view of Lemmas 2.8 and 2.7 this can also be written in the form:

$$[L:K] = [\dot{v}L:\dot{v}K]\cdot[vL^{o}:vK^{o}]\cdot[\bar{L}:\bar{K}]$$

$$= [\dot{v}L:\dot{v}K]\cdot\frac{d_L}{d_K} .$$

Proof: For the moment let us forget the hypothesis that the valuations are p-valuations; consider an arbitrary finite extension $L|K$ of Henselian valued fields. From general valuation theory it is known that the degree formula holds in the form

$$[L:K] = [vL:vK]\cdot[\bar{L}:\bar{K}]\cdot\delta(L|K)$$

where the defect $\delta(L|K)$ is a power of the residue characteristic. Thus Lemma 2.9 says that, in case of p-valuations the defect is $\delta(L|K) = 1$. From general valuation theory there are two criteria known for the defect to be 1:

(i) if the residue characteristic is zero ;

(ii) if the value group vK is isomorphic to $\mathbb{Z}$ and K is of characteristic zero.

Now a p-valuation does in general not satisfy (i) or (ii). On the other hand, if we decompose a p-valuation into its coarse component and its core, then the former satisfies (i) and the latter satisfies (ii); from this Lemma 2.9 follows. The details of the argument are as follows.

Our hypothesis is that K is Henselian; this refers to the given p-valuation v of K. Now the Henselian property is inherited by every coarsening of the given valuation, in particular by the canonical coarse valuation $\dot{v}$ as defined above. Thus K is Henselian also with respect to $\dot{v}$. Since its residue field K^o is of characteristic zero, case (i) applies and yields

$$[L:K] = [\dot{v}L:\dot{v}K]\cdot[L^o:K^o] \ .$$

We know that K^o is p-valued with value group $vK^o \simeq \mathbb{Z}$. Note that K^o is Henselian too; this is readily verified from the Henselian property of K. Hence case (ii) applies to K^o and yields

$$[L^o:K^o] = [vL^o:vK^o]\cdot[\bar{L}:\bar{K}] \quad .$$

Here we have used the fact that K and its core field K^o have the same residue field, and similarly for L. Putting both formulas together we obtain the fundamental degree formula for K and v, in view of Lemma 2.8.

q.e.d.

COROLLARY 2.10 <u>Let</u> $L|K$ <u>be a finite extension of p-valued fields, and assume that</u> K <u>is Henselian. The necessary and sufficient condition for</u> K <u>and</u> L <u>to have the same p-rank is that</u>

$$[L:K] = [\dot{v}L:\dot{v}K] \ .$$

That is, $L|K$ should be fully ramified with respect to the coarse valuation.

This result will be the basis for all what follows.

Remark 2.11: If $L|K$ is an extension of p-valued fields of the same p-rank then the value factor group vL/vK equals the value factor group with respect to the coarse valuation:

$$vL/vK = \dot{v}L/\dot{v}K \quad .$$

This is so because

$$\dot{v}L = vL/vL^{o} \quad , \quad \dot{v}K = vK/vK^{o}$$

and

$$vL^{o} = vK^{o}$$

since

$$[vL^{o}:vK^{o}] = i[vL:vK] = 1 \quad .$$

Therefore, if L and K have the same p-rank then Corollary 2.10 implies

$$[L:K] = [vL:vK]$$

provided that K is Henselian and $L|K$ finite.

The converse, however, is not always true: if $[L:K] = [vL:vK]$ then L and K need not have the same p-rank.

Remark 2.12: Let v be a p-valuation of p-rank d on a field K. Then the coarse valuation $\dot{v}$ induces a place $\varphi:K \to K^{o} \cup \{\infty\}$ onto the core field K^{o} which carries the core valuation, a p-valuation of the same p-rank as v. The completion of K^{o} with respect to the core valuation therefore is isomorphic to some extension field of $\mathbb{Q}_p$ of degree d . Thus K admits a place into some extension of $\mathbb{Q}_p$ of degree d . Conversely, if K admits a place into such an extension of $\mathbb{Q}_p$, using Example 2.5, we obtain a p-valuation on K of p-rank d' where d' divides d .

Therefore, for every d _in_ $\mathbb{N}$, _a field_ K _admits some p-valuation of p-rank dividing_ d _if and only if_ K _admits a place into some extension field of_ $\mathbb{Q}_p$ _of degree_ d .

§ 3. p-adically closed fields

Let K be a p-valued field of p-rank d. We call K p-adically closed if K does not admit any proper p-valued algebraic extension of the same p-rank.

By Zorn's Lemma, if K is p-valued but not necessarily p-adically closed, there exists a maximal p-valued algebraic extension field L of the same p-rank. Any such field L is called a p-adic closure of K. The p-adic closure of a p-valued field K is not necessarily unique. We shall obtain in this section a necessary and sufficient condition for K to admit a unique p-adic closure (up to K-isomorphism). This condition refers to the structure of the value group of the p-valued field K, and it requires that the value group should be a $\mathbb{Z}$-group (see Section 1). If this condition is not satisfied, then we shall show that every p-adic closure of K is generated by radicals $\sqrt[n]{c}$ with $c \in K$ over the Henselization K^h of K. This fact will lead us to the Isomorphism Theorem for p-adic closures (Corollary 3.11).

In the first subsection we will characterize p-adically closed fields (Theorem 3.1). They turn out to be Henselian p-valued fields such that the value group is a $\mathbb{Z}$-group. Most of the theorems in this section including the Isomorphism Theorem already hold for Henselian p-valued fields. Thus we will state and prove them under this more general condition.

3.1 Characterization of p-adically closed fields

THEOREM 3.1 *Let* K *be a p-valued field. Then* K *is p-adically closed if and only if* K *is Henselian and, moreover, its value group* vK *is a* $\mathbb{Z}$*-group. The last condition is equivalent to saying that the coarse value group* $\dot{v}K$ *should be divisible.*

Proof: Clearly, the divisibility property of $\dot{v}K$ is equivalent to the $\mathbb{Z}$-group property of vK.

First suppose that K is Henselian and $\dot{v}K$ is divisible; we claim that K is p-adically closed. Let L be a finite algebraic p-valued extension of K of the same p-rank as K; we have to show $L = K$. By Corollary 2.10 we have

$$[L:K] = [\dot{v}L:\dot{v}K] .$$

Since $\dot{v}K$ is divisible, it does not admit any proper totally ordered group extension of finite index. Hence $\dot{v}L = \dot{v}K$ and we conclude $L = K$, as contended.

Conversely suppose that K is p-adically closed, i.e. K should not admit any proper algebraic p-valued extension of the same p-rank. In particular it follows that the Henselization K^h cannot be a proper extension; thus $K^h = K$ and K is Henselian. We claim that the coarse value group $\dot{v}K$ is divisible. Suppose this were not so; then the following construction would yield a proper algebraic p-valued field extension of the same p-rank, contrary to the hypothesis that K is p-adically closed.

If $\dot{v}K$ is not divisible then there exists a prime number n such that $\dot{v}K$ is not divisible by n. Let $0 \neq c \in K$ such that $\dot{v}c$ is not divisible by n in $\dot{v}K$, and let t be an n-th root of c in the algebraic closure of K. Let us put

$$L = K(t) ;$$

by construction we have

$$t^n = c \in K .$$

Since K is Henselian its p-valuation extends uniquely to a valuation of L and we know that L is p-valued (Lemma 2.7). We claim that

$$[L:K] = n = [\dot{v}L:\dot{v}K] .$$

First we observe that

$$[L:K] = [K(t):K] \leq n$$

since t is a root of the polynomial $X^n - c \in K[X]$ of degree n. Secondly, consider the coarse value group $\dot{v}L$; it contains the element $\dot{v}t$. From $t^n = c$ we conclude $n \cdot \dot{v}t = \dot{v}c$. By construction $\dot{v}c$ is not divisible by n in $\dot{v}K$; hence $\dot{v}t \notin \dot{v}K$. The order of $\dot{v}t$ modulo $\dot{v}K$ is precisely n (since n is a prime number). We see that the factor group $\dot{v}L/\dot{v}K$ contains at least one element of order n and hence

$$n \leq [\dot{v}L:\dot{v}K] .$$

Finally, from the general theory of valued field extensions,

$$[\dot{v}L:\dot{v}K] \leq [L:K] .$$

Combining the above three inequalities, we conclude

$$[L:K] = n = [\dot{v}L:\dot{v}K] \quad ,$$

as contended. Lemma 2.10 shows that $d_L = d_K$.

q.e.d.

In the foregoing construction we may replace the element $c \in K$ by any other element with the same coarse value. For instance we may take πc with $\pi \in K$ a prime element. The resulting field

$$L' = K(t') \quad , \quad t'^n = \pi c$$

has the same properties as proved above for L , i.e.

$$[L':K] = n = [\dot{v}L':\dot{v}K] \quad .$$

In particular it follows that $d_{L'} = d_K$. Now let M be a p-adic closure of L , and M' a p-adic closure of L' . Then M, M' are p-adic closures of K , and we claim that they are <u>not</u> K-isomorphic (as valued fields). For if there would exist a K-isomorphism $M \to M'$ then let us identify $M = M'$ by means of this isomorphism; now both t, t' are contained in M . Let us put

$$a = \frac{t'}{t} \quad .$$

Then

$$a^n = \frac{t'^n}{t^n} = \pi \quad .$$

Hence

$$n \cdot va = v\pi$$

contrary to the fact that π is a prime element not only in K but also in its p-adic closure M .

Thus we see that K admits at least two non-isomorphic p-adic closures, under the hypothesis that K is a p-valued Henselian field and vK not a $\mathbb{Z}$-group. In this statement, the Henselian hypothesis may be dropped. For if K is not Henselian, consider its Henselization K^h. The value groups of K and K^h coincide; hence if vK is not a $\mathbb{Z}$-group the same is true for vK^h. By what has been shown above, K^h is contained in at least two p-adic closures M,M' which are not K^h-isomorphic. If there would exist a K-isomorphism $M \to M'$ (as valued fields) then this would be the identity on K^h due to the universal mapping property of the Henselization. Hence M,M' would be K^h-isomorphic, contradiction.

We have proved:

THEOREM 3.2 *Let* K *be a p-valued field. If* vK *is not a* $\mathbb{Z}$*-group then there exist non-isomorphic p-adic closures of* K. *On the other hand, if* vK *is a* $\mathbb{Z}$*-group then the Henselization* K^h *is p-adically closed by Theorem 3.1, hence* K^h *is the unique p-adic closure of* K.

Thus, the necessary and sufficient condition for K *to admit a unique p-adic closure (up to K-isomorphism), is that the value group* vK *should be a* $\mathbb{Z}$*-group. Or, equivalently, the coarse value group* $\dot{v}K$ *should be divisible.*

Remark 3.3: If vK is not a $\mathbb{Z}$-group then repeated application of the above construction shows that there are *infinitely many* non-isomorphic p-adic closures of K, in fact uncountably many. It is possible to describe the variety of all p-adic closures of K by means of the structure of the coarse value group $\dot{v}K$ and the

Galois cohomology of the core field K^0 ; we shall not go into details here.

THEOREM 3.4 Let L be a p-adically closed field and K a subfield of L , equipped with the valuation induced by L . If K is algebraically closed in L then K is p-adically closed , of the same p-rank as L .

The converse is trivially true: if K is p-adically closed and of the same p-rank as L then K is algebraically closed in L .

Our proof of Theorem 3.4 will proceed in three steps. We shall show successively:

(i) the residue fields of K and of L coincide;

(ii) K contains a prime element of L ;

(iii) the value factor group vL/vK is torsion free.

From (i) and (ii) we deduce that K and L have the same p-rank. In particular it follows that $vL/vK = \dot{v}L/\dot{v}K$ (see Remark 2.11). Hence (iii) can be rephrased as to say that $\dot{v}L/\dot{v}K$ is torsion free. Now $\dot{v}L$ is divisible since L is p-adically closed (Theorem 3.1). It follows that $\dot{v}K$ is divisible. On the other hand, K is Henselian since K is algebraically closed in the Henselian field L . Hence Theorem 3.1 shows that K is p-adically closed.

Thus it remains to prove the assertions (i), (ii) and (iii). For later reference we shall formulate these assertions separately in three lemmas, thereby stating the precise conditions for the validity of those assertions.

LEMMA 3.5 (i) Let L be a Henselian p-valued field, and let $q = p^{f_L}$ denote the order of its residue field $\bar{L}$. Let $K \subset L$ be a subfield.

The polynomial $X^q - X$ splits completely in L, and its roots form a representative set for the residue field $\bar{L}$. By construction, this representative set is algebraic over K.

Consequently, if K is algebraically closed in L then K contains a representative set for $\bar{L}$ and hence $\bar{K} = \bar{L}$.

The proof is immediate from Hensel's Lemma: each $\bar{a} \in \bar{L}$ is a simple root of the polynomial $X^q - X \in \bar{L}[X]$; hence there exists one and only one root $a \in L$ of $X^q - X$ whose residue class is the given $\bar{a}$.

The set of roots of $X^q - X$ is called the Teichmüller representative set of the Henselian p-valued field L.

Before stating the next lemma, let us explain the notion Eisenstein polynomial for a p-valued field L. Let $e = e_L$ denote the p-ramification index of L. An Eisenstein polynomial is of degree e, its highest coefficient is 1, all other coefficients are divisible by p in $\mathcal{O}_L$, and the constant coefficient is exactly divisible by p. Thus an Eisenstein polynomial $f(X) \in L[X]$ is of the form

$$f(X) = X^e - p \cdot g(X)$$

$$g(X) = b_o + b_1 X + \ldots + b_{e-1} X^{e-1}$$

$$vb_i \geq 0 \qquad , \qquad 0 \leq i \leq e-1$$

$$vb_o = 0 .$$

If this Eisenstein polynomial $f(X)$ has a root $\theta \in L$ then a straightforward computation shows that

$$v(\theta^e) = e \cdot v\theta = vp$$

and hence θ is a prime element of L. Every prime element $\pi \in L$ is a root of a suitable Eisenstein polynomial. For we have $\pi^e = pu$ where $vu = 0$, and hence we may take $f(X) = X^e - pu$. The following lemma says that, under certain conditions, an Eisenstein polynomial can be constructed with coefficients in a subfield $K \subset L$, and admitting a root in L.

LEMMA 3.5 (ii) _Let_ L _be a Henselian p-valued field. Let_ $K \subset L$ _be a subfield and assume_ $\bar{K} = \bar{L}$.

Given any prime element π _of_ L, _there exists an Eisenstein polynomial_ $f(X)$ _for_ L _whose coefficients are contained in_ K, _and which satisfies the Newton condition_

$$vf(\pi) > 2 \cdot vf'(\pi) \quad .$$

Hence by Newton's Lemma, $f(X)$ _admits a root_ $\theta \in L$. _By construction_ θ _is a prime element of_ L _which is algebraic over_ K.

Consequently, if K _is algebraically closed in_ L _then_ $\theta \in K$ _and hence_ K _contains a prime element of_ L.

Proof: Since $\bar{K} = \bar{L}$ the field K contains a representative set for the residue field $\bar{L}$. Let R be such a representative set. In choosing R we take care that the zero residue class is represented by the zero element $0 \in K$. Hence if $a \in R$ and $a \neq 0$ then $va = 0$; this will be of importance at a certain stage in the following proof. (If K is algebraically closed in L then we may take for R the Teichmüller representative set, as defined in Lemma 3.5 (i).)

Let $e = e_L$ denote the p-ramification index of L . We identify $v\pi = 1$, so that $\mathbb{Z}$ becomes a convex subgroup of vL; we then have

$$vp = e \quad .$$

Every integer $m \geq 0$ is uniquely representable in the form

$$m = i + ke \qquad \text{with} \quad 0 \leq i < e \ , \ 0 \leq k \ .$$

Let us put

$$\omega_m = \pi^i p^k \quad .$$

Then

$$v\omega_m = m \quad .$$

Hence the monomials $\omega_o = 1$, $\omega_1, \omega_2, \ldots$ are representatives of the non-negative values in $\mathbb{Z}$. Therefore every element $a \in \mathcal{O}_L$ admits an expansion of the form

$$a = a_o + a_1\omega_1 + a_2\omega_2 + \ldots$$

with coefficients a_i from our representative set R . The meaning of this expansion is that the sum of the first s terms is an approximant for a , with remainder of value $\geq s$:

$$v(a - \sum_{o \leq m < s} a_m\omega_m) \geq s \quad ,$$

for each integer $s > 0$. The coefficients $a_m \in R$ are determined successively: a_o is the representative of $\bar{a}$, then a_1 is the representative of $\overline{(a-a_o)/\omega_1}$, then a_2 is the representative of $\overline{(a - a_o - a_1\omega_1)/\omega_2}$, and so on.

Let us consider the expansion for the element $a = \pi^e/p$:

$$\frac{\pi^e}{p} = a_o + a_1\omega_1 + a_2\omega_2 + \dots$$

Since $v(\pi^e/p) = 0$ we have

$$a_o \neq 0 \text{ , i.e. } va_o = 0 \text{ .}$$

Let $s > 0$ be a large integer, to be specified later in the proof. Consider the s-th approximant of π^e/p in the above expansion:

$$\sum_{o \leq m < s} a_m\omega_m = \sum_{\substack{o \leq i < e \\ o \leq k \\ i+ke < s}} a_{i+ke}\pi^i p^k = \sum_{o \leq i < e} b_i \pi^i = g(\pi)$$

where

$$b_i = \sum_{o \leq k < (s-i)/e} a_{i+ke}p^k \quad \text{and} \quad g(X) = \sum_{o \leq i < e} b_i X^i \text{ .}$$

By construction $b_i \in K$ (this is so because $a_{i+ke} \in R \subset K$). Also it is evident from the expansion

$$b_i = \sum_k a_{i+ke}p^k$$

that

$$vb_i \geq 0 \quad , \quad 0 \leq i < s \text{ .}$$

In order to compute vb_i explicitly , we have to determine the smallest exponent k such that p^k actually appears in the above expansion, i.e. such that $a_{i+ke} \neq 0$. Then $va_{i+ke} = 0$ (by our choice of R as explained above) and hence

$vb_i = v(p^k) = ke$. In particular we note that $vb_i \in \mathbb{Z}$ and

$$vb_i \equiv 0 \bmod e \quad , \quad 0 \leq i < s$$

(if $b_i \neq 0$). Since $a_o \neq 0$ we conclude

$$vb_o = 0 \quad .$$

Thus the polynomial

$$f(X) = X^e - p \cdot g(X) \in K[X]$$

is an Eisenstein polynomial for L . We have

$$f(\pi) = p \cdot \left(\frac{\pi^e}{p} - g(\pi) \right) \quad .$$

By construction $g(\pi)$ is the s-th approximant in the expansion for π^e/p ; hence

$$vf(\pi) \geq e + s \quad .$$

In order to estimate $vf'(\pi)$ we write

$$f'(\pi) = e\pi^{e-1} - p \cdot g'(\pi) = e\pi^{e-1} - \sum_{o \leq i < e} pib_i \pi^{i-1} \quad ,$$

$$vf'(\pi) \geq \min_{o<i<e} [v(e)+e-1 \; , \; v(pib_i)+i-1] \quad .$$

We claim that the equality sign holds here. To this end we show that all values appearing in the brackets are mutually different. Indeed we claim they are integers, contained in different residue classes modulo e :

$$v(e) + e-1 \equiv e-1 \bmod e$$

$$v(pib_i) + i-1 \equiv i-1 \bmod e \qquad \text{if } b_i \neq 0 \quad .$$

To see this we recall that $vb_i \equiv 0 \bmod e$ as observed above already. Moreover, for every integer $z \neq 0$ (in particular for $z = e$ and $z = pi$) the value is $vz \equiv 0 \bmod e$; for if p^r is the highest p-power dividing z then $vz = v(p^r) = re$.

We have now shown that

$$vf'(\pi) = \min_{0<i<e} [v(e)+e-1 \ , \ v(pib_i)+i-1] \leq ve + e-1 \ .$$

Therefore, if s is large then

$$vf(\pi) \geq e+s > 2\cdot(ve + e-1) \geq 2\cdot vf'(\pi) \ ;$$

the precise condition for s is

$$s \geq 2\cdot ve + e-1 \ .$$

For such s we see that $f(X)$ satisfies the Newton condition with respect to π .

q.e.d.

LEMMA 3.5 (iii) Let L be a Henselian p-valued field and K a subfield of L . Assume that K and L have the same p-rank.

Let $\alpha \in vL$ be a torsion element modulo vK, and suppose that $n \in \mathbb{N}$ is such that $n\alpha \in vK$. Then there exists $t \in L$ such that $vt = \alpha$ and $t^n \in K$. That is, α is the value of an n-th radical t over K . In particular t is algebraic over K.

Consequently, if K is algebraically closed in L then $t \in K$ and hence $\alpha = vt \in vK$; it follows that vL/vK is torsion free.

Proof: Take $a \in L$ such that $va = \alpha$, and put

$$b = a^n .$$

Then $vb = n \cdot \alpha \in vK$. We are going to show that every $b \in L$ with $vb \in vK$ admits a decomposition in the form

$$b = c \cdot u^n \quad \text{with} \quad c \in K , \quad u \in L , \; vu = 0 .$$

If this is proved then the element

$$t = \frac{a}{u} \in L$$

satisfies the requirements of the lemma: we have

$$vt = va = \alpha ,$$

$$t^n = \frac{b}{u^n} = c \in K .$$

Our assertion about the decomposition of b is the content of part (C) of the following

LEMMA 3.6 Let L be a p-valued Henselian field and $K \subset L$; assume that K and L have the same p-rank.

(A) Let $b \in L$ be such that $vb \in vK$. Given any integer $s > 0$ we may write b in the form

$$b = c \cdot z \quad \text{with} \quad c \in K, \; z \in L, \; v(1-z) \geq s .$$

Hence the multiplicative coset of b modulo $K^{\times}$ contains units z of arbitrary large level s, i.e. $v(1-z) \geq s$.

(B) If $s > 2 \cdot vn$ then every unit $z \in L$ of level s is an n-th power in L. Thus if $v(1-z) \geq s$ then there exists $u \in L$ such *that*

$$z = u^n .$$

u *can be chosen such that*

$$v(1-u) > vn ,$$

and then u *is uniquely determined.*

(C) *Combining* (A) *and* (B) *we see that every* $b \in L$ *with* $vb \in vK$ *admits a decomposition of the form*

$$b = c \cdot u^n \quad \text{with} \quad c \in K,\ u \in L,\ vu = 0 .$$

Hence the multiplicative coset of b *modulo* $K^\times$ *contains* n-th *powers of units, for arbitrary large exponent* n .

Proof: (A) We may replace, if convenient, the given element b by any other element in its coset modulo $K^\times$. Now since $vb \in vK$ there exists $c_o \in K$ such that $vb = vc_o$; after replacing b by bc_o^{-1} we may assume from the start that

$$vb = 0 .$$

We use the expansion

$$b = a_o + a_1\omega_1 + a_2\omega_2 + \ldots$$

as explained in the foregoing proof. The coefficients a_m are contained in a representative set R for the residue field, and R is chosen such that $R \subset K$. The gauge elements ω_m are of the form $\omega_m = \pi^i p^k$ where π is a prime element of L. This time we can choose $\pi \in K$ since the hypotheses of Lemma 3.6 include the assumption that K and L have the same p-rank. It follows that for every $s > 0$ the s-th approximant

$$\sum_{o \le m < s} a_m \omega_m$$

is contained in K. For given s, let c denote this approximant, then

$$v(b-c) \geq s > 0 .$$

Since $vb = 0$ we conclude $vc = 0$ and hence

$$v(\frac{b}{c} - 1) \geq s > 0 .$$

Hence the element $z = bc^{-1}$ is a unit of level s and $b = c \cdot z$ as required.

(B) Consider the polynomial

$$f(X) = X^n - z .$$

We have

$$vf(1) = v(1-z) \geq s$$

since z is supposed to be a unit of level s. On the other hand,

$$vf'(1) = vn .$$

Hence if $s > 2 \cdot vn$ then $vf(1) > 2 \cdot vf'(1)$. Since L is Henselian we conclude from Newton's Lemma that $f(X)$ admits a zero $u \in L$, i.e. $u^n = z$. Moreover, u can be chosen such that $v(1-u) > vf'(1) = vn$, and then u is uniquely determined, according to the uniqueness statement of Newton's Lemma.

q.e.d.

3.2 The Isomorphism Theorem for p-adic closures

LEMMA 3.7 *Let* K *be a p-valued Henselian field and* L *an algebraic extension of* K *of the same p-rank as* K. *If* $vL = vK$ *then* $L = K$.

Proof: Since L is the union of its finite subextensions we may assume that $L|K$ is finite. In this case we know that

$$[L:K] = [vL:vK]$$

from Corollary 2.10 and Remark 2.11. Hence $vL = vK$ implies $L = K$.

q.e.d.

Let us recall the notion of *radical* elements in a field extension $L|K$. An element $0 \neq t \in L$ is called a radical over K if there exists $n \in \mathbb{N}$ such that $t^n \in K$. Thus t is a torsion element modulo $K^\times$, in the multiplicative sense. The radicals of $L|K$ form a multiplicative group T_L which contains $K^\times$.

THEOREM 3.8 *(Radical Structure Theorem) Let* K *be a p-valued Henselian field and* L *an algebraic extension of* K, *of the same p-rank as* K. *Then* $L|K$ *is generated by radicals. Hence if* $T = T_L$ *denotes the radical group of* $L|K$, *we have*

$$L = K(T) .$$

The radical value group vT *equals the full value group of* L :

$$vT = vL .$$

The valuation map $v: T \to vL$ induces an isomorphism of the factor groups:

$$T/K^{\times} \simeq vL/vK \quad .$$

If $[L:K]$ is finite then

$$[T:K^{\times}] = [L:K] \quad .$$

Proof: Since $L|K$ is algebraic the value factor group vL/vK is a torsion group. Hence by Lemma 3.5 (iii) every $\alpha \in vL$ is the value $\alpha = vt$ of some radical $t \in T$; this shows

$$vT = vL \quad .$$

(Note that L is Henselian because it is algebraic over the Henselian field K ; hence Lemma 3.5 (iii) is applicable to the extension $L|K$.) Consider the subfield $L' = K(T)$ of L . Its value group vL' contains $vT = vL$ and hence $vL' = vL$. Applying Lemma 3.7 we conclude

$$L = L' = K(T) \quad .$$

(Again note that L' is Henselian, hence Lemma 3.7 is applicable to the extension $L|L'$.)

Since $vT = vL$ we obtain a surjective homomorphism

$$v: T/K^{\times} \to vL/vK \quad .$$

We claim that this is an isomorphism. Assume that $t \in T$ and $vt \in vK$; we have to show that $t \in K$. Let n denote the order of t modulo $K^{\times}$; then $t^n \in K$. To show that $t \in K$ we may replace t by any other element in its coset modulo $K^{\times}$. (This will not change the order of t modulo $K^{\times}$.) Now we refer to statement (A) of Lemma 3.6 which says that the coset of t modulo $K^{\times}$ contains units of arbitrary high level. Hence we may assume without loss that t itself is a unit of sufficiently high level, more precisely:

$$v(1-t) \geq s \quad \text{with} \quad s > 2 \cdot vn \quad .$$

Let us put

$$t^n = z \in K \; .$$

Since the units of level s form a multiplicative group we have

$$v(1-z) \geq s > 2 \cdot vn \quad .$$

Now we use the fact that K is Henselian. Accordingly we conclude from Newton's Lemma that

$$z = u^n \quad \text{with} \quad u \in K \; .$$

(See statement (B) of Lemma 3.6 , which is now to be applied to K instead of L.) We can choose u such as to satisfy the side condition

$$v(1-u) > vn \quad ,$$

and then u is uniquely determined. Now this uniqueness statement holds not only in K but also in L . In L there is also the element t which satisfies $z = t^n$ and $v(1-t) > 2 \cdot vn \geq vn$. Hence

$$t = u \in K \quad ,$$

as contended .

We have seen that $T/K^\times \simeq vL/vK$. If one of these groups is finite then the other is finite too and it follows $[T:K^\times] = [vL:vK]$. On the other hand, if $[L:K]$ is finite then we know from Corollary 2.10 that $[L:K] = [\dot{v}L:\dot{v}K] = [vL:vK]$ (see also Remark 2.11). We conclude

$$[L:K] = [T:K^\times] .$$

q.e.d.

The Radical Structure Theorem 3.8 will be the main ingredient of our proof of the Isomorphism Theorem. More precisely, we shall use there the following explicit description of the field structure of $L|K$ in terms of radicals, which is an immediate consequence of Theorem 3.8. We keep the hypotheses and notations of Theorem 3.8; for convenience we restrict our discussion to the case of a <u>finite</u> extension $L|K$.

By Theorem 3.8 the radical factor group $T/K^\times$ is finite. Consider a direct decomposition into finite cyclic groups:

$$T/K^\times = C_1 \times C_2 \times \ldots \times C_r .$$

Let n_i be the order of C_i; then

$$n_1 n_2 \ldots n_r = [T:K^\times] = [L:K] ,$$

the latter relation in view of Theorem 3.8. Let $t_i \in T$ be such that t_i generates C_i modulo $K^\times$; then t_i is of order n_i modulo $K^\times$ and hence

$$t_i^{n_i} = c_i \in K \qquad (1 \leq i \leq r) .$$

By Theorem 3.8 we have $L = K(T)$ and hence

$$L = K(t_1,\dots,t_r) \quad .$$

Thus $L|K$ can be generated by finitely many radicals $t_1,\dots,t_r$ such that the product of their radical exponents equals the field degree $[L:K]$.

We claim that the r relations

$$t_i^{n_i} - c_i = 0 \qquad (1 \leq i \leq r)$$

are __defining relations__ for $t_1,\dots,t_r$ over K . In other words: every polynomial relation for $t_1,\dots,t_r$ over K is a consequence of the above r relations. Consider the polynomial ring in r variables

$$K[X_1,\dots,X_r] = K[\underline{X}] \quad .$$

If a polynomial $f(\underline{X}) \in K[\underline{X}]$ satisfies $f(\underline{t}) = f(t_1,\dots,t_r) = 0$ then $f(X)$ is called a __relation__ for $\underline{t}$ over K . These relations form an ideal I of $K[\underline{X}]$, and there is a natural isomorphism of K-algebras

$$K[\underline{X}]/I \simeq L \quad ,$$

which is given by the substitution $\underline{X} \mapsto \underline{t}$. Our claim means that the relation ideal I is generated by the r polynomials $X_i^{n_i} - c_i \in K[\underline{X}]$. To see this let I' denote the ideal generated by these r polynomials; then $I' \subset I$ and we have to show that $I' = I$. Since the i-th polynomial $X_i^{n_i} - c_i$ is of degree n_i in the i-th variable X_i , we have

$$\dim K[\underline{X}]/I' = n_1 n_2 \dots n_r \quad ,$$

the dimension to be understood over K. (This is readily verified by induction on r.) On the other hand we know from Theorem 3.8 that

$$n_1 n_2 \dots n_r = [L:K] = \dim K[\underline{X}]/I .$$

We conclude $I' = I$, as contended.

Let us collect our results in the following

COROLLARY 3.9 In the same situation as in Theorem 3.8, suppose that $L|K$ is finite. Then $L|K$ can be generated by finitely many radicals such that the product of their radical exponents equals the field degree:

$$L = K(t_1,\dots,t_r)$$

$$t_i^{n_i} = c_i \in K \qquad (1 \le i \le r)$$

$$[L:K] = n_1 n_2 \dots n_r .$$

L is K-isomorphic to the polynomial factor algebra

$$L \simeq K[X_1,\dots,X_r]/I$$

where I denotes the ideal generated by the r polynomials

$$X_i^{n_i} - c_i \qquad (1 \le i \le r) .$$

This isomorphism is obtained by the substitution $X_i \mapsto t_i$ $(1 \le i \le r)$.

We are now ready to obtain the Isomorphism Theorem stated in Section 1 from the following more general theorem.

THEOREM 3.10 (Algebraic Embedding Theorem)
Let $L|K$ be an algebraic extension of p-valued fields of the same p-rank, and assume L to be Henselian. Let L' be an

arbitrary Henselian valued field extension of K.

The necessary and sufficient condition for L to be K-isomorphically embeddable into L' (as valued fields) is that

$$K \cap L^n \subset L'^n \qquad \text{for each } n \in \mathbb{N}.$$

That is, if an element $a \in K$ is an n-th power in L then a should also be an n-th power in L'.

Proof: The condition is obviously necessary: if $L \subset L'$ then $L^n \subset L'^n$ for each $n \in \mathbb{N}$. Thus we have to prove the sufficiency of the condition. Let us first consider the case in which the extension $L|K$ satisfies the following two additional hypotheses:

(i) K is Henselian ;

(ii) the field degree $[L:K]$ is finite .

In this case the structure of $L|K$ is described by Corollary 3.9. Namely, $L|K$ can be generated by finitely many radicals such that the product of their radical exponents equals the field degree:

$$L = K(t_1,\ldots,t_r)$$
$$t_i^{n_i} = c_i \in K \qquad (1 \leq i \leq r)$$
$$[L:K] = n_1 n_2 \cdots n_r \quad .$$

We observe that $c_i \in K \cap L^{n_i}$. Hence $c_i \in L'^{n_i}$ by the condition of the theorem. Thus we may write

$$c_i = u_i^{n_i} \quad \text{with} \quad u_i \in L' \qquad (1 \leq i \leq r) \; .$$

We claim that the substitution

$$t_i \mapsto u_i \qquad (1 \leq i \leq r)$$

defines a K-isomorphic embedding

$$\sigma : L \to L' .$$

Consider the polynomial ring $K[X_1,\ldots,X_r] = K[\underline{X}]$. The substitution $\underline{X} \mapsto \underline{u}$ defines a K-homomorphism $K[\underline{X}] \to L'$. Its kernel contains each of the polynomials $X_i^{n_i} - c_i \in K[\underline{X}]$, and hence the kernel contains the ideal I generated by those polynomials. Factoring through I we obtain a K-homomorphism

$$K[\underline{X}]/I \to L' .$$

On the other hand we know from Corollary 3.9 that the substitution $\underline{X} \mapsto \underline{t}$ defines a K-isomorphism

$$K[\underline{X}]/I \simeq L .$$

Putting both maps together we obtain à K-homomorphism

$$\sigma : L \to L'$$

which is necessarily injective since L is a field. By construction we have

$$\sigma t_i = u_i \qquad (1 \leq i \leq r) .$$

It remains to verify that the embedding $\sigma : L \to L'$ is compatible with the given valuations on L and L' , i.e. it should be an embedding as valued fields. This is true since K is assumed to be Henselian. For the Henselian property of K implies that its valuation admits only one prolongation to L . Therefore the given valuation on L coincides with the valuation induced from L' via the K-embedding $\sigma : L \to L'$. Hence indeed, σ is compatible with the valuations.

We have now proved Theorem 3.10 in the special case in which the additional hypotheses (i) and (ii) are satisfied. It will turn out that this special case constitutes the essential part of Theorem 3.10. The following arguments, reducing the general case to this special case, will be of rather formal nature.

First let us remove condition (ii). Thus we consider the case where the base field K is still Henselian, but the algebraic extension $L|K$ may be infinite. L is the union of its finite subextensions $E|K$. The hypotheses of Theorem 3.10 are inherited from $L|K$ to every finite subextension $E|K$. Hence we know from the above that every finite subextension $E|K$ of $L|K$ admits a K-isomorphic embedding $E \to L'$ (as valued fields). From this the embeddability $L \to L'$ can be deduced using standard compactness arguments of Galois theory of infinite algebraic extensions.

We have now proved Theorem 3.10 under the additional hypothesis (i), the extension $L|K$ being finite or infinite. Now consider the general case where (i) may not be satisfied. Hence K may not be Henselian. This can be reduced to the Henselian case as follows.

Since L is Henselian it contains the Henselization K^h of K. More precisely there is a natural K-embedding $K^h \to L$ and we identify K^h with its isomorphic image in L. Similarly $K^h \subset L'$. We now have the following situation:

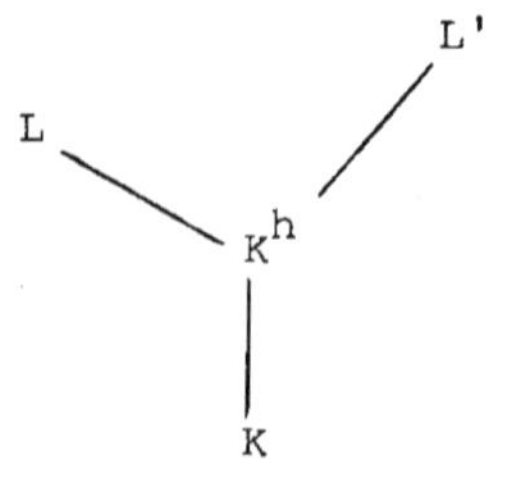

It suffices to construct a K^h-embedding $L \to L'$. Since K^h is Henselian we may apply the foregoing arguments to the extension $L|K^h$ - provided we can show that the condition of the theorem is satisfied by $L|K^h$. Thus we have to show that

$$K \cap L^n \subset L'^n \quad \text{implies} \quad K^h \cap L^n \subset L'^n$$

for each $n \in \mathbb{N}$.

To see this we refer to part (C) of Lemma 3.6, applied to the extension $K^h|K$. Note that K^h and K have the same value group, hence the condition $vb \in vK$ of that lemma is satisfied by every element $0 \neq b \in K^h$. We conclude that every $b \in K^h$ admits a decomposition of the form $b = cu^n$ with $c \in K$ and $u \in K^h$. Hence

$$K^h = K \cdot K^{hn}$$

where $K^{hn} = (K^h)^n$ denotes the n-th power set of the field K^h. Therefore, if $K \cap L^n \subset L'^n$, then $K^h \cap L^n = K \cdot K^{hn} \cap L^n =$
$= (K \cap L^n) \cdot K^{hn} \subset L'^n \cdot L'^n = L'^n$.

q.e.d.

COROLLARY 3.11 (Isomorphism Theorem for algebraic extensions) *Let K be a p-valued field and L, L' two Henselian p-valued algebraic extensions of the same p-rank as K.*

The necessary and sufficient condition for L, L' to be K-isomorphic (as valued fields) is that

$$K \cap L^n = K \cap L'^n \qquad \text{for each } n \in \mathbb{N}.$$

For by Theorem 3.10 this condition expresses the fact that each of the fields L, L' is K-isomorphic to a subfield of the other. Since they are algebraic over K this implies $L \simeq L'$.

The statement of Corollary 3.11 holds in particular if L, L' are two p-adic closures of K, and we obtain the Isomorphism Theorem for p-adic closures as stated in Section 1.

The next structure theorem refers to a p-adic closure of K, say M. From Theorem 3.1 we know that the coarse value group $\dot{v}M$ is divisible; in fact $\dot{v}M$ is the divisible hull of $\dot{v}K$. We may write this in the form

$$\dot{v}M = \mathbb{Q} \otimes \dot{v}K ,$$

the tensor product to be understood over $\mathbb{Z}$. For the value factor group we conclude:

$$vM/vK = \dot{v}M/\dot{v}K = \mathbb{Q}/\mathbb{Z} \otimes \dot{v}K .$$

The group on the right hand side is canonically attached to vK. We have $\mathbb{Q}/\mathbb{Z} \otimes \dot{v}K = 0$ if and only if $\dot{v}K$ is divisible, i.e. vK is a $\mathbb{Z}$-group. The structure of $\mathbb{Q}/\mathbb{Z} \otimes \dot{v}K$ can easily be described: as a divisible torsion group, it is isomorphic to the direct sum of groups of the form $\mathbb{Q}_l/\mathbb{Z}_l$, the l-primary part of $\mathbb{Q}/\mathbb{Z}$, where l is a prime number. For given l, the multiplicity with which $\mathbb{Q}_l/\mathbb{Z}_l$ appears in such direct sum decompositions is counted by the dimension $\dim \dot{v}K/l$, which is to be understood over the prime field $\mathbb{Z}/l$ of l elements. This dimension may be finite or infinite. Note that $\dot{v}K = vK/\mathbb{Z}$, hence $\dim \dot{v}K/l = \dim vK/l - 1$.

THEOREM 3.12 (*Subfield Structure of p-adic closures*)
Let K *be a Henselian p-valued field and* $\dot{v}K$ *its coarse value group. Let us put*

$$\Gamma = \mathbb{Q}/\mathbb{Z} \otimes \dot{v}K .$$

Let M *be a p-adic closure of* K .

The subextensions $L|K$ *of* $M|K$ *correspond* 1-1 *to the subgroups* H *of* Γ. *More precisely, there is a natural lattice isomorphism from the lattice of all subextensions of* $M|K$ *to the lattice of all subgroups of* Γ . *This is obtained by attaching to* $L|K$ *its value factor group* $vL/vK \subset vM/vK$ *and using the natural identification* $vM/vK = \Gamma$ *as explained above. If* $L|K$ *is finite then the corresponding subgroup* $H \subset \Gamma$ *is finite too and* $[L:K] = |H|$.

It follows that the lattice structure of subextensions of $M|K$ is uniquely determined by K , and does not depend on the choice of the p-adic closure M of K .

Proof: It remains to prove the following two statements:

(i) *Each subextension* $L|K$ *of* $M|K$ *is uniquely determined by its value factor group* $vL/vK \subset \Gamma$.

(ii) *Every subgroup* $H \subset \Gamma$ *is the value factor group* $H = vL/vK$ *of a suitable subextension* $L|K$ *of* $M|K$.

Let T_L denote the radical group of the field extension $L|K$. By Theorem 3.8 we have a natural isomorphism

$$T_L/K^{\times} \simeq vL/vK \quad ,$$

defined by the valuation v on L . If we regard L as a subfield of M we have $T_L \subset T_M$, and the above isomorphism is induced by the isomorphism

$$T_M/K^\times \simeq vM/vK = \Gamma \quad .$$

In other words: Γ is naturally isomorphic to $T_M/K^\times$; in this isomorphism the value factor group vL/vK of a subextension corresponds to the radical factor group T_L/K of the same subextension. Hence the statements (i) and (ii) are translated, via the isomorphism $T_M/K^\times \simeq \Gamma$, into the following two statements:

(i') Each subextension $L|K$ of $M|K$ is uniquely determined by its radical factor group $T_L/K^\times$.

(ii') Every subgroup $T/K^\times$ of $T_M/K^\times$ is the radical factor group of a suitable subextension $L|K$ of $M|K$.

Statement (i') is already contained in Theorem 3.8, namely $L = K(T_L)$. Statement (ii') is proved as follows:

Let T be an arbitrary subgroup of T_M such that $K^\times \subset T$. Let $L = K(T)$ be the field extension generated by T . Then clearly

$$K^\times \subset T \subset T_L$$

and we claim $T = T_L$. Every $a \in L$, and hence every $a \in T_L$ is contained in some extension $K(T')$ with $T' \subset T$ and T'/K finite. Hence after replacing T by T' we may assume that $T/K^\times$ is finite. Consider a direct decomposition into cyclic groups:

$$T/K^\times = C_1 \times C_2 \times \dots \times C_r \quad .$$

Let n_i be the order of C_i, and let $t_i \in T$ be such that t_i generates C_i modulo $K^\times$. Then we have

$$L = K(T) = K(t_1,t_2,\dots,t_r) \quad .$$

Each t_i is a root of an equation of degree n_i over K , hence we conclude

$$[L:K] \leq n_1 n_2 \ldots n_r = [T:K^\times] \leq [T_L:K^\times] .$$

On the other hand, by Theorem 3.8 we have

$$[L:K] = [T_L:K^\times] .$$

Therefore $[T:K^\times] = [T_L:K^\times]$ and $T = T_L$.

q.e.d.

§ 4. The General Embedding Theorem

The Embedding Theorem 3.10 will now be generalized to the case of transcendental extensions $L|K$. However, in this case the Henselian property of the receiving field L' will not be sufficient. For instance if $L'|K$ is algebraic and $L|K$ is not algebraic then we cannot expect to obtain an embedding $L \to L'$. Hence we have to add a condition which will ensure that L' is "large enough" to receive an isomorphic image of L . This will be achieved by assuming L' to be κ-saturated for a sufficiently large cardinal κ , namely $\kappa > |L|$. This notion of saturation is taken from model theory; for a general definition see [S], §16. The concept of κ-saturated structures refers to a given language in which the theory under consideration is presented; in our context we always mean the language of valued fields. The structures under consideration are fields together with a valuation ring.

A valued field K is called κ-saturated (where κ is an uncountable cardinal) if for every set Φ of formulas in one variable, say X , and with parameters from K (i.e. involving constants for elements of K) the following holds: if Φ has cardinality less than κ and every finite set of formulas $\varphi_1(X),\ldots,\varphi_n(X)$ from Φ is realized by some $y \in K$, then there is an element $y \in K$ realizing all formulas $\varphi(X)$ from Φ simultaneously.

The main application of the General Embedding Theorem will be in Section 7.1 . For the convenience of the reader we will single out there what we have really used here from the saturation property, and then give a selfcontained proof.

In addition to the saturation property we shall assume that L' is not only Henselian but also p-adically closed. This will ensure that the value group vL' is sufficiently large compared to the value group of L. (See also Remark 4.12)

This being said the General Embedding Theorem can now be formulated as follows:

THEOREM 4.1 (General Embedding Theorem) Let $L|K$ be an extension of p-valued fields of the same p-rank, and assume L to be Henselian. Let L' be a p-adically closed extension of K (as a valued field), such that L' is sufficiently high saturated. Explicitly, L' should be κ-saturated for some cardinal $\kappa > |L|$.

The necessary and sufficient condition for L to be K-isomorphically embeddable into L' (as valued field) is that

$$K \cap L^n \subset L'^n \qquad \text{for each } n \in \mathbb{N}.$$

4.1 Reductions of the theorem

Let $\tilde{K}$ denote the algebraic closure of K in L. Then

$$K \cap L^n = K \cap \tilde{K}^n .$$

Thus the condition of Theorem 4.1 concerns the algebraic part $\tilde{K}|K$ only, rather than the whole extension $L|K$. Using the Algebraic Embedding Theorem 3.10 we conclude that $\tilde{K}$ is K-embeddable into L'. Identifying $\tilde{K}$ with its image in L' we now have the following situation:

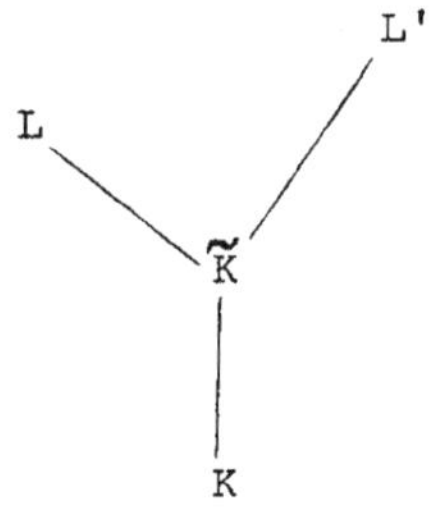

It suffices to construct a $\widetilde{K}$-embedding of L into L' . Hence after replacing K by $\widetilde{K}$ we may assume that K is algebraically closed in L , i.e. that $L|K$ is a *regular* field extension. Note that if $L|K$ is regular then

$$K \cap L^n = K^n$$

and hence the condition of Theorem 4.1 is vacuous. Therefore in the regular case, Theorem 4.1 asserts that $L|K$ can always be embedded into L' . This is an important special case of Theorem 4.1 which we want to state in a separate theorem. Before doing so, however, let us pause a moment to give a criterion for regularity of $L|K$ in terms of the value groups.

LEMMA 4.2 *Let* $L|K$ *be an extension of p-valued fields, of the same p-rank. Suppose* L *to be Henselian. Then* K *is algebraically closed in* L *if and only if* K *is Henselian and, moreover, the value factor group* vL/vK *is torsion free*.

Proof: Since L is Henselian it contains the Henselization of K, i.e.

$$K \subset K^h \subset L .$$

If K is algebraically closed in L we conclude $K = K^h$, i.e.

K is Henselian. Also, we have already proved in Lemma 3.5 (iii) that vL/vK is torsion free.

Conversely, suppose K to be Henselian and vL/vK torsion free. Let $\tilde{K}$ denote the algebraic closure of K in L. Then

$$K \subset \tilde{K} \subset L$$

$$vK \subset v\tilde{K} \subset vL \ .$$

The factor group $v\tilde{K}/vK$ is a torsion group and contained in the torsion free group vL/vK. Hence $v\tilde{K} = vK$. Lemma 3.7 now shows that $\tilde{K} = K$. (Note that K is supposed to be Henselian, hence Lemma 3.7 is applicable to the extension $\tilde{K}|K$.)

q.e.d.

Referring to the discussion preceeding Lemma 4.2 we see that the proof of the General Embedding Theorem is reduced to the following "regular case".

THEOREM 4.3 (Embedding Theorem for regular extensions)
Let $L|K$ be an extension of p-valued fields of the same p-rank. Assume that K is Henselian and vL/vK is torsion free. (In view of Lemma 4.2 this means that $L^h|K$ is a regular extension.)

Then L is K-isomorphically embeddable into every p-adically closed extension L' of K which is sufficiently high saturated. In fact, L' should be κ-saturated for some cardinal $\kappa > |L|$.

Remark 4.4: The hypotheses of Theorem 4.3 do not include the assumption that L is Henselian, because this is not necessary for the validity of the theorem. On the other hand, for the proof *we may assume, whenever convenient, that* L *too is Henselian*. For if L should not be Henselian then we can replace L by its Henselization L^h ; any embedding $L^h \to L'$ will induce an embedding $L \to L'$.

We start the proof of Theorem 4.3 with the reduction to the case where $L = K(x)$ is a rational function field in one variable. This reduction will be of rather formal nature. The core of proof will then be the discussion of embeddings of rational function fields.

Remark 4.5: *Suppose that Theorem 4.3 is valid whenever* $L|K$ *is of transcendence degree 1. Then Theorem 4.3 is valid in general*.

Proof: Let $L|K$ and $L'|K$ satisfy the hypotheses of Theorem 4.3; we have to show that L admits a K-isomorphic embedding into L' (as valued fields). In view of Remark 4.4 we may assume that L is Henselian. Let $\mathcal{T}$ denote the system of all subextensions $K_1|K$ of $L|K$ such that K_1 is algebraically closed in L . Every K_1 in $\mathcal{T}$ is Henselian. Consider the K-embeddings $K_1 \to L'$ which are defined on fields K_1 in $\mathcal{T}$. By Zorn's Lemma there exists a *maximal* such embedding $K_1 \to L'$ with K_1 in $\mathcal{T}$. We have to show that $K_1 = L$. If this were not so then there would exist $x \in L$, $x \notin K_1$. Let L_1 denote the algebraic closure of $K_1(x)$ in L . Then L_1 is in $\mathcal{T}$ and $L_1|K_1$ is of transcendence degree 1. After identifying K_1 with its isomorphic image in L' we now have the following situation:

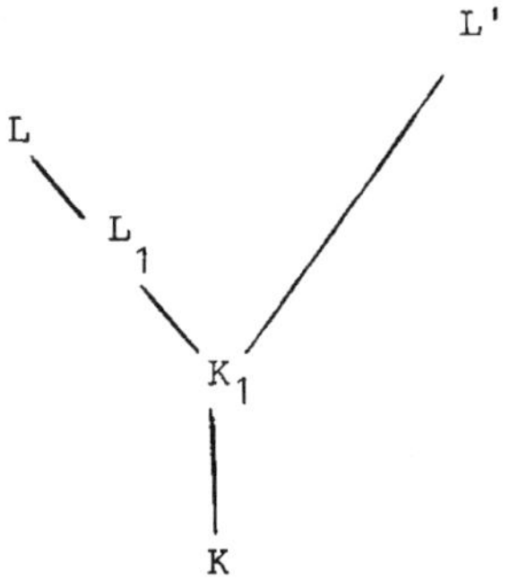

With respect to the extension $L_1|K_1$ this is precisely the situation of the Embedding Theorem 4.3. Hence, if we know that Theorem 4.3 is valid for the case of transcendence degree 1 , then we infer that there exists a K_1-embedding of L_1 into L'. In other words: the given embedding $K_1 \to L'$ is extended to an embedding $L_1 \to L'$; since L_1 is in $\mathcal{T}$ this contradicts the maximality property of $K_1 \to L'$.

q.e.d.

Remark 4.6: In the situation of Theorem 4.3 suppose that every finitely generated subextension of $L|K$ can be K-embedded into L' . Then the whole field L can be K-embedded into L' .

This is an easy consequence of the κ-saturation property of L', for $\kappa > |L|$. Notice however: we do not claim that every K-embedding into L' of a finitely generated subextension can be extended to a K-embedding of the whole field L; this is not true in general.

By the above remarks the proof of Theorem 4.3 is reduced to the case where $L|K$ is finitely generated and of transcendence degree 1 . In order to deal with this case we need the following two lemmas:

LEMMA 4.7 *Let* $L|K$ *be an extension of valued fields, finitely generated and of transcendence degree* 1 *. Suppose the value factor group* vL/vK *is torsion free. Then there exists* $x \in L$, *transcendental over* K , *such that the value group of the rational function field* $K(x)$ *equals the value group of the whole field:*

$$vK(x) = vL \ .$$

More explicitly:

Case A: if $vL = vK$ *then every transcendental element* $x \in L$ *over* K *satisfies the above condition.*

Case B: if $vL \neq vK$ *then the factor group* vL/vK *is infinite cyclic. If* $x \in L$ *is chosen such that its value* $vx = \xi$ *generates* vL *modulo* vK *then* x *is transcendental over* K , *and it satisfies the above condition.*

LEMMA 4.8 *Let* $L = K(x)$ *be a rational function field in one variable. Suppose* L *carries a valuation such that the value* $vx = \xi$ *is of infinite order modulo* vK . *Then for any nonzero polynomial*

$$f(x) = \sum_i a_i x^i \in K[x]$$

we have

$$vf(x) = \min_i [va_i + i\xi] .$$

The value group vL *is the direct sum of* vK *and the cyclic group generated by* ξ :

$$vL = vK \oplus \mathbb{Z}\cdot\xi \ .$$

We shall prove Lemma 4.8 first.

Proof of Lemma 4.8: If $f(x)$ is as in the lemma then, clearly,

$$vf(x) \geq \min_i [va_i + i\xi] \ .$$

In order to show that the equality sign holds we have to show that all the appearing terms $va_i + i\xi$ are mutually different (if $a_i \neq 0$). Indeed: if $i < j$ and $va_i + i\xi = va_j + j\xi$ then $(j-i)\xi = va_i - va_j \in vK$, contrary to the fact that ξ is of infinite order modulo vK. Thus we have shown that

$$vf(x) = \min_i [va_i + i\xi] \ .$$

In particular we see that $vf(x) \in vK + \mathbb{Z}\cdot\xi$. Every element in $L = K(x)$ is the quotient $f(x)/g(x)$ of two polynomials; hence its value is $vf(x) - vg(x) \in vK + \mathbb{Z}\cdot\xi$. We conclude

$$vL = vK + \mathbb{Z}\cdot\xi .$$

The sum is direct since ξ is of infinite order modulo vK.

q.e.d.

Proof of Lemma 4.7: In case A, if $vL = vK$, there is nothing to prove. So let us consider the case B, where $vL \neq vK$. Let $\xi_o \in vL$, $\xi_o \notin vK$. Then ξ_o is of infinite order modulo vK (since vL/vK is torsion free). Choose $x_o \in L$ such that $vx_o = \xi_o$. Then x_o is transcendental over K (since otherwise vx_o would be of finite order modulo vK). Hence for the rational function field $L_o = K(x_o)$ we have the situation of Lemma 4.8. Therefore

$$vL_o = vK + \mathbb{Z}\cdot\xi_o \ .$$

In particular we see that vL_o/vK is infinite cyclic.

Now we use the fact that $L|K$ is finitely generated and of transcendence degree 1 . This implies that L is of finite degree over its rational subfield L_o . In particular it follows that vL/vL_o is finite. We conclude that vL/vK contains the infinite cyclic group vL_o/vK as a subgroup of finite index. Since vL/vK is torsion free this implies that vL/vK itself is infinite cyclic.

Accordingly let $\xi \in vL$ be a generator modulo vK . Then

$$vL = vK + \mathbb{Z} \cdot \xi \ .$$

If $x \in L$ is chosen such that $vx = \xi$ we see that $vK(x)$ contains vK and ξ , hence

$$vK(x) = vL \ .$$

q.e.d.

Let us return to the proof of Theorem 4.3 . We have already reduced the proof to the case where $L|K$ is finitely generated and of transcendence degree 1 . In this case choose $x \in L$ according to Lemma 4.7 , so that x is transcendental over K and

$$vK(x) = vL \ .$$

Since we are dealing with p-valued fields of the same p-rank, this implies that L is contained in the Henselization $K(x)^h$ of $K(x)$. To see this, consider L^h as an algebraic extension of $K(x)^h$, both p-valued and of the same p-rank. Their value groups coincide:

$$vL^h = vL = vK(x) = vK(x)^h \ .$$

Now Lemma 3.7 shows that $K(x)^h$ does not admit any proper algebraic p-valued extension of the same p-rank and the same value group. We conclude $L^h = K(x)^h$ and hence

$$L \subset K(x)^h$$

as contended.

Now let L' be an extension of K as required in Theorem 4.3 . If we can construct a K-embedding $K(x) \to L'$ then this extends uniquely to an embedding $K(x)^h \to L'$ (since L' is Henselian); since $L \subset K(x)^h$ we obtain a K-embedding $L \to L'$. We conclude:

Remark 4.9: *In order to prove Theorem 4.3 it is sufficient to prove it in the case where* $L = K(x)$ *is a rational function field in one variable and one of the following cases applies:*

Case A: $vL = vK$

Case B: $vx = \xi$ *is of infinite order modulo* vK *and hence* $vL = vK \oplus \mathbb{Z} \cdot \xi$ *(by Lemma 4.8).*

4.2 Proof of the rational case

PROPOSITION 4.10A *Let* K *be a Henselian p-valued field. Suppose that the rational function field* $L = K(x)$ *is equipped with a p-valuation extending the given valuation of* K *, such that* L *and* K *have the same p-rank and that*

$$vL = vK .$$

Let L' *be a valued extension of* K *which is* κ*-saturated for* some cardinal $\kappa > |K|$. Then L' contains an element y such that

$$v(x-a) = v(y-a) \qquad \text{for all} \quad a \in K \ .$$

That is, x and y have the same distance to each element $a \in K$. If this is so then for each polynomial $f(x) \in K[x]$ we have

$$vf(x) = vf(y) \ .$$

Therefore the substitution $x \mapsto y$ yields a value preserving K-isomorphic embedding of $L = K(x)$ into L'.

Proof: Clearly, L is an immediate extension of K, in the sense of valuation theory.

Now let $a \in K$. The value $v(x-a)$ equals the value of some element $b \in K$:

$$v(x-a) = vb \quad , \quad v\left(\frac{x-a}{b}\right) = 0 \ .$$

The residue class of $\frac{x-a}{b}$ equals the residue class of some element $c \in K$:

$$v\left(\frac{x-a}{b} - c\right) > 0 \quad , \quad v(x - (a+bc)) > vb = v(x-a) \ .$$

Hence putting $y = a + bc$ we see that the following statement holds:

Given $a \in K$ there exists $y \in K$ such that $v(x-y) > v(x-a)$ and therefore $v(x-a) = v(y-a)$.

Now let $a_1, \ldots, a_r$ be finitely many elements in K. Among the r values $v(x-a_i)$ consider the largest one, say $v(x-a)$, and choose $y \in K$ with respect to a according to the above statement. We obtain:

$$v(x-y) > v(x-a) \geq v(x-a_i)$$

and therefore

$$v(x-a_i) = v(y-a_i) .$$

Hence the following statement holds:

<u>Given finitely many</u> $a_1,\dots,a_r \in K$ <u>there exists</u> $y \in K$ <u>such that</u> $v(x-a_i) = v(y-a_i)$ <u>for</u> $1 \leq i \leq r$.

At this point we use the saturation property of L' . The above statement shows that finitely many of the conditions

$$v(x-a) = v(y-a) \qquad (a \in K)$$

can always be realized by some $y \in L'$ (and even $y \in K$). The number of those conditions equals the cardinality $|K|$. Therefore, since L' is saturated with respect to some cardinal $\kappa > |K|$ we conclude that there exists $y \in L'$ satisfying simultaneously all of the above conditions, for all $a \in K$. (Note that $vL = vK$, hence the values $v(x-a)$ are contained in $vK \subset vL'$. Therefore the conditions

$$v(x-a) = v(y-a) \qquad (a \in K)$$

are meaningful, if regarded in the value group vL'.)

Let us choose $y \in L'$ as above. We have to show the relation

$$vf(x) = vf(y)$$

for every polynomial $0 \neq f(x) \in K[x]$. Certainly this is true for polynomials of degree 0 (i.e. constants). A polynomial of degree 1 is (up to a constant factor) of the form $x-a$ with $a \in K$; hence the choice of y implies the validity of the above relation for polynomials of degree 1 .

Suppose the above relation were not always true; then let $f(x) \in K[x]$ be a <u>counterexample of smallest degree</u>:

$$vf(x) \neq vf(y)$$

$$n = \deg f(x) \text{ minimal .}$$

By what we have said above, $n > 1$. If $f(x) = g(x) \cdot h(x)$ would be reducible then its factors g,h have smaller degree and the relations

$$vg(x) = vg(y) \quad , \quad vh(x) = vh(y)$$

imply $vf(x) = vf(y)$. We conclude that $f(x)$ is <u>irreducible</u>. Let $F = K[x]/f(x)$ be the corresponding residue field. We have

$$[F:K] = n > 1 \ .$$

Every element of F can be uniquely represented by a polynomial of degree $< n$ in $K[x]$. For the moment let us identify the elements of F with their unique representatives. Then F is identified with the vector space of polynomials of degree $< n$:

$$F = K + K \cdot x + \ldots + K \cdot x^{n-1} \quad .$$

This is an identification of K-vector spaces. The field multiplication in F is not the ordinary polynomial multiplication; let us denote the field multiplication by $g(x) * h(x)$ for $g(x)$, $h(x) \in F$. By definition $g(x) * h(x) = r(x)$ is the unique polynomial of degree $< n$ which is obtained from the product $g(x) \cdot h(x)$ by Euclidean division with $f(x)$:

$$g(x) \cdot h(x) = s(x) \cdot f(x) + r(x) \quad ,$$

with $s(x) \in K[x]$. Counting degrees, we find that $s(x) \in F$.

By our above identification we have

$$K \subset F \subset L$$

as K-vector spaces. The given valuation v of L induces on F a function

$$v: F \to vK$$

which satisfies all properties of valuations except possibly the multiplication rule for field multiplication; by this we mean the rule

$$v(g(x)*h(x)) = vg(x) + vh(x)$$

for $0 \neq g(x), h(x) \in F$. We claim that this rule is also satisfied. For suppose that there exist $g(x), h(x) \in F$ not satisfying this rule, i.e.

$$vr(x) \neq vg(x) + vh(x)$$

where $r(x) = g(x)*h(x)$, as above. From

$$s(x) \cdot f(x) = g(x) \cdot h(x) - r(x)$$

we conclude

$$vf(x) = -vs(x) + \min[vg(x)+vh(x), vr(x)] .$$

This shows that the value $vf(x)$ can be expressed by the values of polynomials of degree $< n$. The latter values are preserved by the substitution $x \mapsto y$ (because $f(x)$ was supposed to be a counterexample of minimal degree). So we have

$$vg(x) = vg(y) \quad , \quad vh(x) = vh(y)$$

$$vr(x) = vr(y) \quad , \quad vs(x) = vs(y) .$$

By means of the above formula we conclude $vf(x) = vf(y)$, a contradiction.

Thus we see that the function

$$v: F \to vK$$

is in fact a valuation of the field F. In this way F becomes a valued field extension of K. The value group vF is equal to vK by the very definition. In particular, a prime element $\pi \in K$ is also a prime element in F. The residue field $\bar{F}$ coincides with $\bar{K}$. For if $vg(x) \geq 0$ then there exists $c \in K$ with $v(g(x) - c) > 0$; this is true in $L = K(x)$ and hence in F because addition and subtraction in F coincide with addition and subtraction in L. We conclude that F is p-valued, of the same p-rank as K, and $vF = vK$.

At this point we use the fact that K is Henselian. Hence K does not admit any proper algebraic field extension of the same p-rank and the same value group (Lemma 3.7). It follows $F = K$, contrary to the fact $[F:K] = n > 1$ as observed earlier already.

q.e.d.

In the following proposition it is convenient to use the sign function $\operatorname{sgn}(\gamma)$, which is defined in every totally ordered group Γ such that

$$\operatorname{sgn}(\gamma) = \begin{cases} +1 & \text{if } \gamma > 0 \\ 0 & \text{if } \gamma = 0 \\ -1 & \text{if } \gamma < 0 \end{cases}.$$

PROPOSITION 4.10B *Let K be a p-valued field. Suppose that the rational function field $L = K(x)$ is equipped with a p-valuation extending the given valuation of K, such that $vx = \xi$ is of infinite order modulo vK and hence (by Lemma 4.8)*

$$vL = vK \oplus \mathbb{Z}\cdot\xi \quad .$$

Let L' be a p-adically closed extension of K which is κ-saturated for some cardinal $\kappa > |K|$. Then L' contains an element y whose value $vy = \eta$ satisfies

$$\mathrm{sgn}(\xi-\alpha) = \mathrm{sgn}(\eta-\alpha) \qquad \text{for all } \alpha \in \mathbb{Q}\otimes vK \; .$$

That is, ξ and η determine the same Dedekind cut in the divisible hull $\mathbb{Q}\otimes vK$ of vK. The substitution $\xi \mapsto \eta$ defines an order preserving embedding of the value group vL into vL'. Let us identify vL with its isomorphic image so that

$$vL \subset vL'$$
$$vx = vy \; .$$

Then for each polynomial $f(x) \in K[x]$ we have

$$vf(x) = vf(y) \; .$$

Therefore the substitution $x \mapsto y$ yields a value preserving K-isomorphic embedding of $L = K(x)$ into L'.

We identify $vp = 1 \in vK$. If $\pi \in K$ is a prime element then $v\pi = \frac{1}{e}$ where e is the p-ramification index of K. The additive group $\mathbb{Z}\cdot\frac{1}{e}$ is a convex subgroup of vK. Hence $\mathbb{Q}$ is a convex subgroup of the divisible hull $\mathbb{Q}\otimes vK$ of vK. Similarly we see that $\mathbb{Q}$ is a convex subgroup of $\mathbb{Q}\otimes vL$.

Since ξ is of infinite order modulo vK it is not contained in $\mathbb{Q} \otimes vK$. Hence for any $\alpha \in \mathbb{Q} \otimes vK$ we have either $\alpha < \xi$ or $\xi < \alpha$. Let us discuss the first case:

$$\alpha < \xi .$$

Since $\mathbb{Q}$ is convex in $\mathbb{Q} \otimes vL$ we clearly have

$$\alpha < \alpha + q < \xi \qquad \text{for each } q \in \mathbb{Q},\ q > 0 .$$

Now we claim that

$$\alpha + q \in vL' \qquad \text{for suitable } q \in \mathbb{Q},\ q > 0 .$$

To see this we use the hypothesis that L' is p-adically closed. By Theorem 3.1 this implies that the coarse value group $\dot{v}L'$ is divisible. We have $\dot{v}L' = vL'/\mathbb{Z}\cdot\frac{1}{e'}$, where e' is the p-ramification index of L'. (Note that our hypothesis does not include the assumption that K and L' have the same p-rank, nor that they have the same p-ramification index). Hence the divisibility of $\dot{v}L'$ can be expressed by the formula

$$vL' = \mathbb{Z}\cdot\frac{1}{e'} + n\cdot vL' \qquad \text{for each } n \in \mathbb{N} .$$

This implies

$$\frac{1}{n}\cdot vL' = \mathbb{Z}\cdot\frac{1}{e'n} + vL' \subset \mathbb{Q} + vL' .$$

Since $n \in \mathbb{N}$ is arbitrary we conclude

$$\mathbb{Q} \otimes vL' = \mathbb{Q} + vL' .$$

Therefore every element in $\mathbb{Q} \otimes vL'$, in particular our element α above, admits a representation in the form

$$\alpha = q_o + \eta_o \qquad \text{with} \quad q_o \in \mathbb{Q}\ ,\ \eta_o \in vL' \ .$$

Choose an integer $z \in \mathbb{Z}$ such that $z > q_o$. Then

$$\alpha + (z-q_o) = (z+\eta_o) \in vL' \quad ,$$

as contended. We have thus proved the following statement:

Given $\alpha \in \mathbb{Q} \otimes vK$ with $\alpha < \xi$, there exists $\eta \in vL' \cap \mathbb{Q} \otimes vK$ such that $\alpha < \eta < \xi$ and hence $\text{sgn}(\xi-\alpha) = \text{sgn}(\eta-\alpha)$.

The case $\xi < \alpha$ is discussed similarly. From these two cases we get at once the following generalization.

Given finitely many $\alpha_1,\ldots,\alpha_r \in \mathbb{Q} \otimes vK$ there exists $\eta \in vL' \cap \mathbb{Q} \otimes vK$ such that $\text{sgn}(\xi-\alpha_i) = \text{sgn}(\eta-\alpha_i)$ for $1 \le i \le r$.

At this point we use the saturation property of L' . The above statement shows that finitely many of the conditions

$$\text{sgn}(\xi-\alpha) = \text{sgn}(\eta-\alpha) \qquad (\alpha \in \mathbb{Q} \otimes vK)$$

can always be realized by some element $\eta \in vL'$ (and even $\eta \in vL' \cap \mathbb{Q} \otimes vK$). The number of these conditions equals the cardinality of $\mathbb{Q} \otimes vK$, which in turn is $\le$ the cardinality $|K|$. Therefore, since L' is saturated with respect to some cardinal $\kappa > |K|$ we conclude that there exists $\eta \in vL'$ satisfying simultaneously all of the above conditions, for all $\alpha \in \mathbb{Q} \otimes vK$.

In the following, let us choose some $\eta \in vL'$ according to the above conditions. Note that $\eta \notin \mathbb{Q} \otimes vK$ since otherwise we could take $\alpha = \eta$ in the above conditions, which would give

$\operatorname{sgn}(\eta-\alpha) = 0$ while $\operatorname{sgn}(\xi-\alpha) \neq 0$, a contradiction. Hence η is of infinite order modulo vK .

Let us recall that the value group vL is the direct sum

$$vL = \mathbb{Z}\cdot\xi + vK .$$

Accordingly the substitution $\xi \mapsto \eta$ defines an embedding of groups

$$vL \to vL' .$$

Explicitly , this map is given by

$$n\xi + \gamma \mapsto n\eta + \gamma \qquad (\gamma \in vK,\ n \in \mathbb{Z}) .$$

By the choice of η this map is compatible with the order structures in vL and vL' respectively.

Now let us identify vL with its order-isomorphic image in vL'; this identification means that the valuation v on L is replaced by an equivalent valuation, having the same valuation ring. Note that by this identification the values $vc = \gamma$ for $c \in K$ do not change. We have

$$vx = \xi = \eta = vy$$

if $y \in L'$ is chosen such that its value is η .

The element y is transcendental over K (because its value is, by construction, of infinite order modulo vK). Hence the substitution $x \mapsto y$ defines a K-embedding of the rational function field:

$$L = K(x) \to L' .$$

We claim that this embedding is compatible with the valuations defined on L and L' . Let

$$f(x) = \sum_i a_i x^i$$

be a nonzero polynomial in K[x]. By Lemma 4.8 we have

$$vf(x) = \min_i [va_i + i\xi] .$$

On the other hand, Lemma 4.8 may also be applied to K(y) instead of K(x); we obtain

$$vf(y) = \min_i [va_i + i\eta] .$$

The identification $\xi = \eta$ yields

$$vf(x) = vf(y) .$$

Hence, indeed, the substitution $x \mapsto y$ defines a value preserving K-embedding of $L = K(x)$ into L' .

q.e.d.

Remark 4.11: In the situation of Theorem 4.3 , suppose in addition that vL = vK . Then the conclusion of the theorem holds for any sufficiently high saturated Henselian extension L' of K; it is not necessary that L' is p-adically closed.

This is so because the crucial Proposition 4.10A , which is applicable in this case, does not require the receiving field L' to be p-adically closed. Also, the reduction process from the general case of Theorem 4.3 to the special case of Proposition 4.10A does not use the p-adic closure property of L' (the Henselian property of L' is used in the proof of Remark 4.9).

Note that the reduction process preserves the extra hypothesis $vL = vK$ so that indeed, after the reduction is achieved, Proposition 4.10A is applicable.

In the general case, if $vL \neq vK$, then there are counterexamples showing that in Theorem 4.3 the p-adic closure property of the receiving field L' cannot be replaced by the Henselian property. However in analyzing the proof of Proposition 4.10B we see that the p-adic closure property of L' is used only to deduce that

$$\mathbb{Q} \otimes vL' = \mathbb{Q} + vL' .$$

This is a condition for the value group vL' only; there is no condition for the residue field of L'. Note that $\mathbb{Q}$ is embedded into the divisible hull $\mathbb{Q} \otimes vL'$ by identifying $vp = 1$. We obtain:

Remark 4.12: *The conclusion of Theorem 4.3 holds for any sufficiently high saturated Henselian extension L' of K whose value group vL' satisfies*

$$\mathbb{Q} \otimes vL' = \mathbb{Q}\cdot vp + vL' .$$

More generally, it suffices that there is an intermediate field L" of K and L' satisfying

$$\mathbb{Q} \otimes vL'' = \mathbb{Q}\cdot vp + vL'' .$$

If, for instance, K is p-adically closed, L" = K satisfies this condition.

§ 5. Model theory of p-adically closed fields

The basic notions from model theory used in this section can be found in books like [B-S], [C–K] or [S].

The <u>language of valued fields</u> is the first order language whose vocabulary contains, besides the logical symbols, function symbols for the field operations (addition, multiplication, subtraction, division), constants for the neutral elements 0,1 and one <u>additional predicate</u> which in a valued field is interpreted to denote the <u>valuation ring</u>. It is clear that the axioms of valued fields can be formulated in this language.

We do not include a predicate for the value group in our language, nor a function symbol for the valuation. This is not necessary because any first order statement about the value group, as a totally ordered group, can be expressed in terms of the field operations and the valuation ring.

For a given prime number p and p-rank d, the defining property

$$d = \dim \mathcal{O}/p$$

of p-valued fields of p-rank d can be expressed in the language of valued fields. <u>Hence the class of p-valued fields of p-rank d is axiomatizable in the language of valued fields</u>.

The axiom "$d = \dim \mathcal{O}/p$" contains existential quantifiers. For it says that "there exist elements $u_1,\ldots,u_d$ in $\mathcal{O}$ which form a $\mathbb{Z}/p$-basis of $\mathcal{O}/p$". If we insist upon <u>universal axioms</u>

then we have to modify the language of valued fields by adding d new constants to denote the elements of a basis of $\mathcal{O}/p$. In this modified language, the axiom can now be stated in the form: "$u_1, \ldots, u_d$ are elements of $\mathcal{O}$ and form a $\mathbb{Z}/p$-basis of $\mathcal{O}/p$". This axiom can be formulated without any existential quantifier since the base field has only p elements.

With respect to this modified vocabulary, every substructure of a p-valued field of p-rank d is again a p-valued field of p-rank d.
(Without the modification, in the original vocabulary of valued fields, this would not be true: a substructure of a p-valued field of rank d would then be a p-valued field of rank $\leq d$.)

This remark will be of importance in discussing the problem of quantifier elimination, where we shall have to modify the language of valued fields in the indicated manner. Of course, if $d = 1$ then we need not modify the language since then the constant symbol for the field element 1 can be used as symbol for a basis of $\mathcal{O}/p$. (Also in general we could have been a little more economical by using only $d - 1$ new constants, which together with the field element 1 should denote a basis of $\mathcal{O}/p$.)

As defined at the beginning of Section 3, a p-valued field K of p-rank d is called p-adically closed if it does not admit any proper algebraic p-valued extension field of the same p-rank d. This definition is a priori not a statement within the language of valued fields. However, we proved in Section 3

(Theorem 3.1) that p-adically closed fields can be characterized as p-valued fields which are Henselian and whose value groups are $\mathbb{Z}$-groups. Let us discuss these last properties and show that they can be formulated by axioms within the language of valued fields; it will then follow from Theorem 3.1 that <u>the class of p-adically closed fields of p-rank d is axiomatizable in the language of valued fields</u>.

The value group vK is called a <u>$\mathbb{Z}$-group</u> if it is elementary equivalent, in the language of ordered groups, to the group $\mathbb{Z}$ of integers. This can also be expressed in the following manner. As above, π denotes a prime element of K and we identify $v\pi = 1$ so that $\mathbb{Z}$ becomes a convex subgroup of vK. Now vK is a $\mathbb{Z}$-group if and only if for each $n \in \mathbb{N}$ the following statement holds: every $\alpha \in vK$ is congruent modulo n to one of the integers $0,1,\ldots,n-1$. That is, there should exist $\xi \in vK$ such that

$$\alpha = i + n\xi \qquad \text{with} \quad i \in \{0,1,\ldots,n-1\} \quad .$$

If $\alpha = va$ with $a \in K$, then this condition means there should exist $x \in K$ such that

$$va = v(\pi^i x^n) \qquad \text{with} \quad i \in \{0,1,\ldots,n-1\} \quad .$$

In other words:

$$\frac{a}{\pi^i x^n} \in \mathcal{O} \quad \text{and} \quad \frac{\pi^i x^n}{a} \in \mathcal{O} \quad .$$

Thus we see that, for given $n \in \mathbb{N}$, the above statement can be formulated in the language of valued fields. (Observe that the property of π to be a prime element can be expressed in the language of valued fields.) Therefore, <u>the class of p-valued fields of p-rank d whose value group is a $\mathbb{Z}$-group is axiomatizable in the language of valued fields</u>. There are infinitely many axioms, one for each $n \in \mathbb{N}$. As we saw in Theorem 3.2, these fields are precisely those p-valued fields of p-rank d which admit a unique p-adic closure.

From the discussion at the beginning of Section 2.2 we can see that <u>the class of Henselian valued fields is axiomatizable within the language of valued fields</u>. Indeed, Hensel's Lemma (or Newton's Lemma) for polynomials of degree n can be formulated within the language of valued fields. Thus we have infinitely many axioms, one for each degree n .

From Theorem 4.3 - the regular case of the General Embedding Theorem - we will now deduce the <u>Model Completeness Theorem</u>:

THEOREM 5.1 <u>The theory of p-adically closed fields of (fixed) p-rank</u> d <u>is model complete</u>,
i.e. <u>if</u> $L|K$ <u>is an extension of p-adically closed fields of p-rank</u> d , <u>then</u> L <u>is an elementary extension of</u> K.

Let us recall that an extension $L|K$ (of valued fields) is called an <u>elementary extension</u> if every elementary sentence (about valued fields) with parameters from K holds in L if and only if it holds in K .

<u>Proof</u>: By Robinson's Test (cf. [C-K], Prop. 3.17) it suffices to show for each such extension $L|K$ that any existential sentence φ with parameters from K which holds in L also holds in K .

By general model theory K admits elementary extensions of arbitrary high saturation. Choosing an elementary extension L' of K of sufficiently high saturation and observing that a p-adically closed field K is algebraically closed in every p-valued extension field L of the same p-rank d, we infer from Theorem 4.3 that L can be K-isomorphically embedded into L'.

From the existential character of φ we now get that φ holds in L'. Thus by the choice of L' it also holds in K.

q.e.d.

THEOREM 5.2 <u>The theory of p-adically closed fields of p-rank</u> d <u>is decidable, i.e. there is an effective procedure to determine for each sentence</u> φ <u>in the language of valued fields, whether</u> φ <u>holds in all p-adically closed fields of p-rank</u> d, <u>or not</u>.

<u>Proof</u>: By Gödel's Completeness Theorem, a sentence φ holds in each p-adically closed field of p-rank d if and only if φ is deducible by the usual first order calculus of logic from the axioms characterizing those fields. Since the axioms are effective, we thus obtain an effective procedure to produce all valid sentences. It remains to find an effective procedure to produce the non-valid sentences, i.e. those sentences φ which are wrong in at least one p-adically closed field L of p-rank d.

Assume φ is wrong in L. Let K be the algebraic closure of $\mathbb{Q}$ in L. From Theorem 3.4 we infer that K is p-adically closed of p-rank d with respect to the valuation induced from L.

By 5.1 L is an elementary extension of K. Hence φ is also wrong in K. Denote by Q_p the Henselization of $\mathbb{Q}$ with respect to the p-adic valuation. We may assume $Q_p \subset K$. Clearly, K is a finite algebraic extension of Q_p. Let $K \simeq Q_p[X]/\hat{g}(X)$ where $\hat{g}(X) \in Q_p[X]$ is irreducible and monic. Choosing $g(X) \in \mathbb{Q}[X]$ close enough to $\hat{g}(X)$ (with respect to the p-adic valuation of Q_p applied to the coefficients) we can guarantee that $g(X)$ is irreducible and has a zero in K too. Say $g(x) = 0$ and $K = Q_p(x)$. Now K is the Henselization of the finite number field $F = \mathbb{Q}(x)$. Moreover, the p-adic valuation of $\mathbb{Q}$ extends uniquely to F. Indeed, since Q_p is p-adically closed of p-rank 1, we get

$$[K:Q_p] = [vK:vQ_p]\cdot[\bar{K}:\bar{Q}_p] = e_K \cdot f_K = d .$$

Now from the equality

$$e_F \cdot f_F = e_K \cdot f_K = [K:Q_p] = [F:\mathbb{Q}]$$

it follows by general valuation theory that F has only one extension of the p-adic valuation of $\mathbb{Q}$.

We now claim that the sentence $\neg\varphi$ is deducible from the axioms of p-adically closed fields of p-rank d together with the additional axiom

$$\exists x \quad g(x) = 0 .$$

Let K' be any model of this axiom system. Then K' is p-adically closed of p-rank d, containing the number field F. Hence K' also contains the Henselization K of F (with respect to the unique extension of the p-adic valuation of $\mathbb{Q}$). Since K

and K' are of the same p-rank d and $\neg\varphi$ holds in K , by 5.1 it also holds in K'. Hence $\neg\varphi$ is deducible from the above mentioned axioms. By this we obtain an effective method for producing all sentences φ wrong in K .

The proof of this theorem is now completed by the observation that there are only finitely many choices for the field K in the above argument. More precisely, there are only finitely many non-isomorphic algebraic Henselian extensions K of $\mathbb{Q}$ of p-rank d . This can be easily deduced from well-known facts in number theory. We refer the reader to [N], Chapter V .

q.e.d.

From the above proof we also obtain

COROLLARY 5.3 *Let* L *be the completion of a finite algebraic p-valued number field of p-rank* d . *Then* Th(L) (*i.e. the set of sentences true in* L) *can be axiomatized using the axioms for p-adically closed fields of p-rank* d *and an axiom of the type*

$$\exists x \; g(x) = 0$$

where g(X) *is suitably chosen from* $\mathbb{Z}[X]$. *In particular*, Th(L) *is decidable*.

Since there is (up to isomorphism) only one unramified extension of $\mathbb{Q}_p$ of p-rank d (see [N], Chapter V), Theorem 5.1 also implies

COROLLARY 5.4 *The theory of p-adically closed unramified (i.e. with p-ramification* $e=1$) *fields of p-rank* d *is complete and thus also decidable*.

Using now the full strength of the General Embedding Theorem we obtain

THEOREM 5.5 (*Elementary Equivalence Theorem*)

Let K *be a p-valued field and let* L, L' *be p-adically closed extensions of* K , *of the same p-rank as* K . *Suppose that*

$$K \cap L^n = K \cap L'^n \qquad \textit{for each} \quad n \in \mathbb{N} .$$

Then L *and* L' *are elementary equivalent over* K .

The converse is trivially true: if L and L' are elementary equivalent over K then an element $a \in K$ is an n-th power in L if and only if a is an n-th power in L' .

Proof: Let $\tilde{K}$ be the algebraic closure of K in L . Then $K \cap L^n = K \cap \tilde{K}^n$. Similarly for the algebraic closure $\tilde{K}'$ of K in L' . Hence the hypothesis of the theorem implies that

$$K \cap \tilde{K}^n = K \cap \tilde{K}'^n \qquad \text{for each} \quad n \in \mathbb{N} .$$

The Isomorphism Theorem 3.11 now shows that $\tilde{K}$ and $\tilde{K}'$ are K-isomorphic. Identifying $\tilde{K} = \tilde{K}'$ we obtain the following situation:

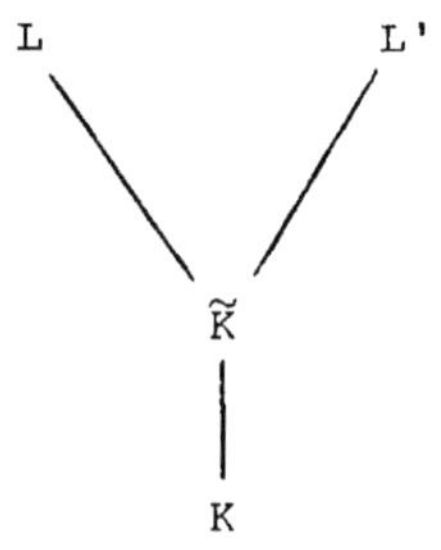

It suffices to show that L and L' are elementary equivalent over $\tilde{K}$ (rather than over K). Hence after replacing K by $\tilde{K}$ we may assume that K is algebraically closed in L and in L' .

Applying Theorem 3.4 we conclude that K is p-adically closed. (Note that step (i) and (ii) of the proof of Theorem 3.4 are trivially satisfied here.) From Theorem 5.1 we infer that L is an elementary extension of K, and so is L'. Hence L and L' are elementary equivalent over K.

q.e.d.

Recall that on page 86 we modified the language of valued fields (by adding certain constants) in order to be able to axiomatize the class of p-valued fields of p-rank d by universal axioms. In particular, with respect to this modified language, every substructure of a p-valued field of p-rank d is again a p-valued field of p-rank d.

It is now clear that Theorem 5.5 just expresses <u>substructure completeness</u> (see [S] §13) for the theory of p-adically closed fields of p-rank d axiomatized in the modified language and extended by unary predicates P_n $(n \in \mathbb{N})$ and the defining axioms

$$\forall x (P_n(x) \leftrightarrow \exists y (x = y^n)) .$$

Since for a first order theory, substructure completeness equals Elimination of Quantifiers we thus generalize a result of MacIntyre [MI]:

THEOREM 5.6 (<u>Quantifier Elimination Theorem</u>)

<u>In the modified language, extended by unary predicates</u> P_n $(n \in \mathbb{N})$, <u>the theory of p-adically closed fields</u> K <u>of p-rank</u> d <u>for which</u> P_n <u>is interpreted by the power set</u> K^n, <u>admits elimination of quantifiers</u>.

§ 6. Formally p-adic fields

6.1 Characterization of formally p-adic fields

A field K is said to be *formally p-adic* if K admits at least one p-valuation. If we specify the p-rank d of the required p-valuation then K is called *formally p-adic of p-rank d*. Similarly we may specify the p-ramification index e and (or) the residue degree f; note that $d = e \cdot f$. Our aim is to characterize formally p-adic fields by means of field theoretic properties.

For a moment, let us look at the analogous situation in the case of formally real fields. A field is said to be formally real if it admits at least one ordering. Formally real fields are characterized by the property that -1 is not a sum of squares in the field. We see that in the real case, the square operator X^2 plays a fundamental role. One of its essential properties is that in any ordered field, squares are always positive.

Returning to the p-adic case we are looking for an operator $\gamma(X)$ which can serve in the same way as the square operator does in the real case. Let e and f be given natural numbers. We put $q = p^f$ and

$$\gamma(X) = \frac{1}{p}\cdot[(X^q-X)-(X^q-X)^{-1}]^{-e} .$$

$\gamma(X)$ is called the *p-adic Kochen operator of type* (e,f). If we want to indicate the dependence on e and f we write $\gamma_{e,f}(X)$. In most cases however, we prefer to write simply $\gamma(X)$; it will be clear from the context which parameters e and f we are referring to.

The following lemma represents the p-adic analog to the fact that, in the real case, squares are always positive. In the p-adic case the γ-operator serves to deal with those p-valuations whose p-ramification index is $\le e$ and whose residue degree divides f. It will be convenient to use the following terminology.

Let v be a p-valuation of a field. If the p-ramification index of v is $\le e$ and if the residue degree of v divides f then v is said to be of type (e,f).

Let K be a field of characteristic zero. K is said to be formally p-adic of type (e,f) if K admits at least one p-valuation of type (e,f).

LEMMA 6.1 Let K be a field of characteristic zero and v a valuation of K. Then v is a p-valuation of type (e,f) if and only if $v(p) > 0$ and $v(\gamma K) \ge 0$, where $\gamma(X)$ is the p-adic Kochen operator of type (e,f).

As to the notation, γK denotes the set of all elements $\gamma(a)$ with $a \in K$. The relation $v(\gamma K) \ge 0$ means that $\gamma K \subset \mathcal{O}_v$, the valuation ring of v in K. In particular this implies that $\gamma(a) \neq \infty$ for every $a \in K$.

Before turning to the proof of Lemma 6.1 let us first generalize the problem such as to include also the "relative" case, which will be of interest later on. In the relative case a p-valued field k is given, serving as base field. Let K be an extension field of k. Then K is said to be formally p-adic over k if K admits a p-valuation v extending the given p-valuation of k .

Suppose this to be the case. Then the p-ramification index e_v of v is a multiple of the p-ramification index e_k of the given p-valuation of k (see Lemma 2.7). Let us write $e_v = e_v' \cdot e_k$ with some natural number e_v'. Then e_v' is the (relative) initial index of v over k . We have

$$e_v' = v(\pi)$$

when $\pi \in k$ is a prime element for the given p-valuation of k . In this formula we have normalized the p-valuation v of K such that $v(\Pi) = 1$, where Π denotes a prime element for v in K . Without that normalization, the above formula reads:

$$e_v' \cdot v(\Pi) = v(\pi) \quad .$$

In a similar way the relative residue degree of v over k is defined. Namely, the residue degree f_v of v is a multiple of the residue degree f_k of the given p-valuation of k . Let us write $f_v = f_v' \cdot f_k$. Then f_v' is the relative residue degree of v . We have

$$f_v' = [\bar{K}_v : \bar{k}] \quad ,$$

the degree of the respective residue fields.

Now let e and f be given natural numbers. v is said to be of relative type (e,f) over k if $e_v' \le e$ and $f_v' | f$.

The extension field K of k is said to be formally p-adic over k , of relative type (e,f) if K admits a p-valuation v over k , of relative type (e,f).

In order to describe formally p-adic field extensions of relative type (e,f), the following relative Kochen operator will be used. Let π denote a prime element of the p-valued base field k. Let $q_k = p^{f_k}$ denote the order of the residue field $\bar{k}$ and put $q = q_k^f$. Then

$$\gamma(X) = \frac{1}{\pi} \cdot [(X^q - X) - (X^q - X)^{-1}]^{-e}$$

is called the <u>π-adic Kochen operator over</u> k <u>of type</u> (e,f). In this context, whenever we speak of the π-adic Kochen operator over k, it is always understood that π is a fixed prime element of k, chosen once for all during the investigation. The results obtained will not depend on the choice of π.

LEMMA 6.2 <u>As explained above, we work over a p-valued base field k. Let K be an extension field of k, and v a valuation of K extending the given p-valuation of k. A necessary and sufficient condition for v to be a p-valuation over k, of relative type (e,f) is that</u>

$$v(\gamma K) \geq 0$$

<u>where $\gamma(X)$ denotes the π-adic Kochen operator of type (e,f) over k.</u>

If $k = \mathbb{Q}$ with its unique p-valuation, and if $\pi = p$ then the above "relative" notions coincide with the corresponding "absolute" notions introduced before. Hence Lemma 6.2 is the "relativistic" generalization of Lemma 6.1.

Let us put

$$\gamma(X) = \frac{1}{\pi} \cdot \beta(X)^e$$

where

$$\beta(X) = [(X^q-X)-(X^q-X)^{-1}]^{-1}$$

$$= \frac{X^q - X}{(X^q-X)^2 - 1}$$

$$= \frac{1}{2}\left[\frac{1}{X^q-X-1} + \frac{1}{X^q-X+1}\right] .$$

As a preparation to the proof of Lemma 6.2 let us give the following list of values $v(\beta a)$ for $a \in K$.

LEMMA 6.3

(i) <u>If</u> $v(a) > 0$ <u>then</u> $v(a^q-a) = v(a) > 0$ <u>and hence</u> $v(\beta a) = v(a) > 0$.

(ii) <u>If</u> $v(a) < 0$ <u>then</u> $v(a^q-a) = q\cdot v(a) < 0$ <u>and hence</u> $v(\beta a) = -q\cdot v(a) > 0$.

(iii) <u>If</u> $v(a) = 0$ <u>and</u> $v(a^q-a) > 0$ <u>then</u> $v(\beta a) = v(a^q-a) > 0$.

(iv) <u>If</u> $v(a) = 0$ <u>and</u> $v(a^q-a) = 0$ <u>then</u> $v(\beta a) \leq 0$.

The verification is immediate from the definition of the β-operator. Case (iv) includes the case when $\beta(a) = \infty$, i.e. $a^q-a = \pm 1$.

<u>Proof</u> of Lemma 6.2: Assume that v is a p-valuation over k , of relative type (e,f) over k . In particular, the relative residue degree f_v' divides f . Hence the residue field $\bar{K}_v$ is contained in the field with degree f over $\bar{k}$, i.e. in the field with $q = q_k^f$ elements. It follows that the polynomial $X^q - X$ vanishes on $\bar{K}_v$; this means that $v(a^q-a) > 0$ for every $a \in K$ with $v(a) \geq 0$. We conclude that case (iv) in Lemma 6.3 does not occur in K . Hence for every $a \in K$ one of the cases

(i)-(iii) applies and therefore $v(\beta a) > 0$. This implies $v(\beta a) \geq v(\Pi) = 1$, where Π denotes a prime element of v in K . Consequently

$$v(\beta(a)^e) = e\cdot v(\beta a) \geq e \geq e_v' = v(\pi)$$

We conclude

$$v(\gamma a) = v(\frac{\beta(a)^e}{\pi}) \geq 0 .$$

This holds for every $a \in K$, hence $v(\gamma K) \geq 0$.

Conversely assume that $v(\gamma K) \geq 0$. Recall that v is supposed to be an extension of the given p-adic valuation of the base field k . Therefore $v(\pi) > 0$. Let $a \in K$ such that $0 < v(a) \leq v(\pi)$. From case (i) of Lemma 6.3 we infer that $v(\beta a) = v(a)$ and therefore $v(\gamma a) = e\cdot v(a) - v(\pi)$. Since $v(\gamma a) \geq 0$ by assumption, we conclude that

$$\frac{v(\pi)}{e} \leq v(a) .$$

This holds for every value $v(a) \in vK$ which is contained in the interval $0 < v(a) \leq v(\pi)$. Now let us divide this interval into e subintervals as follows:

$$0 < \frac{v(\pi)}{e} < 2\cdot\frac{v(\pi)}{e} < \ldots < v(\pi) .$$

Each of these (upper half closed) subintervals contains at most one element of the value group vK . For suppose that, say,

$$(i-1)\cdot\frac{v(\pi)}{e} < v(a_1) < v(a_2) \leq i\cdot\frac{v(\pi)}{e} ,$$

then for $a = a_2 a_1^{-1}$ we have

$$0 < v(a) = v(a_2) - v(a_1) < \frac{v(\pi)}{e} ,$$

contrary to what has been shown above. Thus indeed each of the e subintervals contains at most one element $v(a) \in vK$, hence there are at most e elements of vK in the full interval $0 < v(a) \leq v(\pi)$.

We conclude, firstly that there exists a smallest positive element in the value group vK, say $v(\Pi)$ with Π a prime element of v in K . Let us identify $v(\Pi) = 1$. Secondly, we conclude that at most e of the elements $1,2,3,...$ are $\leq v(\pi)$ in the value group vK . It follows $v(\pi) = e'$ with some natural number $e' \leq e$. By definition, e' is the relative initial index of v over k .

Next we show that v is of finite relative residue degree f' over k , and that f' divides f . This means that the residue field $\bar{K}_v$ should be contained in the field with $q = q_k^f$ elements. In other words: the polynomials $X^q - X$ should vanish on $\bar{K}_v$. Let $a \in K$ and $v(a) = 0$. We claim that case (iv) of Lemma 6.3 does not apply. For if $v(\beta a) \leq 0$ then $v(\gamma a) = e \cdot v(\beta a) - v(\pi) < 0$ contrary to the hypothesis that $v(\gamma K) \geq 0$. (Recall that $v(\pi) > 0$.) We conclude that case (iii) of Lemma 6.3 applies and hence $v(a^q - a) > 0$. This means that the residue class $\bar{a} \in \bar{K}_v$ satisfies $\bar{a}^q - \bar{a} = 0$. Hence indeed, the polynomial $X^q - X$ vanishes on $\bar{K}_v$.

We have shown that the valuation v has finite initial index $e' \leq e$ over k, and finite relative residue degree $f' | f$. Hence v is a p-valuation, and of relative type (e,f) over k (see Lemma 2.7).

q.e.d.

Now let $\mathfrak{o}$ denote the valuation ring of the p-valued field k. Consider the subring of K which is generated by $\mathfrak{o}$ and by the set $\gamma(K) \setminus \{\infty\}$. This ring should be denoted by $\mathfrak{o}[\gamma(K) \setminus \{\infty\}]$ but we shall write $\mathfrak{o}[\gamma K]$ in order to simplify notation. If K is formally p-adic over k, of relative type (e,f) then γK does not contain ∞ as follows from Lemma 6.2; hence our notation $\mathfrak{o}[\gamma K]$ is literally correct in this case.

THEOREM 6.4 *Let* K *be an extension field of the p-valued field* k. *Let* $\gamma(X)$ *denote the* π-*adic Kochen operator of type* (e,f) *over* k, *where* π *is a prime element of the p-valued field* k. *A necessary and sufficient condition for* K *to be formally p-adic over* k *of relative type* (e,f) *is that*

$$\frac{1}{\pi} \notin \mathfrak{o}[\gamma K] .$$

That is, π *should not be a unit in* $\mathfrak{o}[\gamma K]$.

The analogy of this criterion to the corresponding criterion in the real case is apparent: in the real case the criterion reads that -1 should not be a linear combination of squares with positive coefficients in the base field. Here in the p-adic case, $\frac{1}{\pi}$ should not be a polynomial in γ's with integral coefficients in the base field.

Proof: Necessity: Suppose K admits a p-valuation v over k, of relative type (e,f). Let $\mathcal{O}_v$ denote the corresponding valuation ring. By Lemma 6.2 we have $\gamma K \subset \mathcal{O}_v$ and hence $\mathfrak{o}[\gamma K] \subset \mathcal{O}_v$. Since π is not a unit in $\mathcal{O}_v$ we conclude that π is not a unit in $\mathfrak{o}[\gamma K]$ either.

Sufficiency: Suppose π is not a unit in $\mathfrak{o}[\gamma K]$. We invoke the general existence theorem for valuations to conclude that there exists a valuation v of K lying above $\mathfrak{o}[\gamma K]$ and centered over π. This means that the valuation ring $\mathcal{O}_v$ contains $\mathfrak{o}[\gamma K]$ and that its maximal ideal contains π. The intersection $P = \mathfrak{M}_v \cap \mathfrak{o}[\gamma K]$ is called the center of v on $\mathfrak{o}[\gamma K]$. Indeed, given an arbitrary prime ideal $\mathfrak{P}$ of $\mathfrak{o}[\gamma K]$ containing π then the general existence theorem provides us with a valuation v of K lying above $\mathfrak{o}[\gamma K]$ and centered precisely at $\mathfrak{P}$.

We claim that v extends the given p-valuation of the base field k; this means $k \cap \mathcal{O}_v = \mathfrak{o}$. Indeed, on the one hand we have $\mathfrak{o} \subset \mathfrak{o}[\gamma K] \subset \mathcal{O}_v$ and hence $\mathfrak{o} \subset k \cap \mathcal{O}_v$. On the other hand let $a \in k \setminus \mathfrak{o}$; then a^{-1} is contained in the maximal ideal $\pi \cdot \mathfrak{o}$ of $\mathfrak{o}$, hence $a^{-1} \in \pi \cdot \mathcal{O}_v$. Since π is contained in the maximal ideal of $\mathcal{O}_v$ we see that a^{-1} is not a unit in $\mathcal{O}_v$ which implies $a \notin \mathcal{O}_v$. Hence $k \cap \mathcal{O}_v = \mathfrak{o}$.

Since $\mathfrak{o}[\gamma K] \subset \mathcal{O}_v$ we have $\gamma(K) \setminus \{\infty\} \subset \mathcal{O}_v$. (See the above definition of $\mathfrak{o}[\gamma K]$.) Hence if $A \subset K$ denotes the subset of those elements $a \in K$ for which $\gamma(a) \neq \infty$ then $\gamma(A) \subset \mathcal{O}_v$. An element

$a \in K$ is contained in A if and only if a is not a zero of the denominator of the Kochen operator $\gamma(X)$, i.e. if and only if $(a^q-a)^2 - 1 \neq 0$. There are at most finitely many elements in K which do not meet this condition; thus A is cofinite in K. On the other hand, since $\gamma(A) \subset \mathcal{O}_v$ we see that A is the fore-image of $\mathcal{O}_v$ under the map $\gamma : K \to K \cup \{\infty\}$. This map is continuous (referring to the topology of K resp. $K \cup \{\infty\}$ as defined by the valuation v). Since $\mathcal{O}_v$ is closed in $K \cup \{\infty\}$ it follows that A is closed in K. Hence A is both cofinite and closed in K; it follows $A = K$. (Note that every non-empty open set in K is infinite since v is not the trivial valuation.) Thus we have seen that

$$\gamma K \subset \mathcal{O}_v .$$

Now we conclude from Lemma 6.2 that v is a p-valuation of K over k, and that v is of relative type (e,f) over k. Consequently K is formally p-adic over k of relative type (e,f).

q.e.d.

Our above proof of Theorem 6.4 yields the following corollary which will be used in Section 6.2 below.

COROLLARY 6.5 In the situation of Theorem 6.4 let v be a valuation of K. A necessary and sufficient condition for v to be a p-valuation over k of relative type (e,f) is that v lies above $\mathfrak{o}[\gamma K]$ and is centered over π.

6.2 The Kochen ring

We continue to discuss the "relative" situation as in the foregoing section. Thus k is a given p-valued field. Let e and f be natural numbers and $\gamma(X)$ the π-adic Kochen operator over k, of type (e,f).

For any extension field K of k the γ-Kochen ring R_γ is defined as a certain subring of K, namely the ring of quotients of the form

$$a = \frac{b}{1+\pi c} \qquad \text{with} \quad b,c \in \mathfrak{o}[\gamma K] \text{ and } \quad 1+\pi c \neq 0 .$$

If we want to indicate which field K we are considering then we write $R_\gamma(K)$. On the other hand, if it is clear from the context which operator $\gamma(X)$ we are referring to then we simply write R.

The quotient field of the Kochen ring is $k(\gamma K)$, the field generated over k by the set $\gamma(K) \setminus \{\infty\}$. Now we refer to Merckel's Lemma (proved in the appendix) which says that $k(\gamma K) = K$. Hence:

LEMMA 6.6 *The γ-Kochen ring R of $K|k$ admits K as its field of quotients*: $\mathrm{Quot}(R) = K$.

Now we claim:

LEMMA 6.7 *If π is a unit in $\mathfrak{o}[\gamma K]$ then* $R = K$.

Proof: If π is a unit in $\mathfrak{o}[\gamma K]$ then $1 + \pi \cdot \mathfrak{o}[\gamma K] = \mathfrak{o}[\gamma K]$. That is, every element in $\mathfrak{o}[\gamma K]$ is of the form $1 + \pi c$ with $c \in \mathfrak{o}[\gamma K]$. Consequently, the definition of the Kochen ring shows that R is the quotient field of $\mathfrak{o}[\gamma K]$, hence $R = K$ by Lemma 6.6.

q.e.d.

THEOREM 6.8 *Suppose that* π *is not a unit in* $\mathfrak{o}[\gamma K]$; *in view of Theorem 6.4 this is equivalent to saying that* K *is formally p-adic over* k , *of relative type* (e,f). *Then:*

(i) π *is not a unit in the* γ-*Kochen ring* R . *Every maximal ideal of* R *contains* π , *and every prime ideal of* R *containing* π *is maximal.*

(ii) *The p-valuations of* $K|k$ *of relative type* (e,f) *can be characterized as being those valuations of* K *which lie above* R *and are centered at some maximal ideal of* R .

Proof: Let P be a maximal ideal of R . Suppose P does not contain π ; then π is a unit modulo P , i.e. there exists $a \in R$ such that $\pi \cdot a \equiv 1 \bmod P$. Writing $a = \frac{b}{1+\pi c}$ with $b,c \in \mathfrak{o}[\gamma K]$ we obtain $1 + \pi(c-b) \equiv 0 \bmod P$. Multiplying by $\frac{1}{1+\pi(c-b)} \in R$ we obtain $1 \equiv 0 \bmod P$ a contradiction. Note that $1 + \pi(c-b) \neq 0$ since π is supposed not to be a unit in $\mathfrak{o}[\gamma K]$; hence indeed $\frac{1}{1+\pi(c-b)} \in R$ by definition of the Kochen ring.

This proves that every maximal ideal P of R contains π . In particular π is not a unit in R .

Now let P be an arbitrary prime ideal of R containing π . By the general existence theorem for valuations there exists a valuation v of K lying above R and centered at P on R . Then v lies above $\mathfrak{o}[\gamma K]$ (since $\mathfrak{o}[\gamma K] \subset R$) and $v(\pi) > 0$ (since $\pi \in P$). From Corollary 6.5 we infer that v is a p-valuation over k , of relative type (e,f).

In particular it follows that the residue field $\bar{K}_v$ is finite. Since R/P is contained in $\bar{K}_v$ and since every subring of a finite field is a field, we conclude that R/P is a field. Consequently P is a maximal ideal of R. At the same time we have proved that every valuation v of K lying above R and centered at some maximal ideal P of R, is a $\underline{p}$-valuation over k of relative type (e,f).

Finally, let v be an arbitrary p-valuation of $K|k$ of relative type (e,f). From Corollary 6.5 we infer that v lies above $\mathcal{O}[\Upsilon K]$, which is to say that $v(b) \geq 0$ for every $b \in \mathcal{O}[\Upsilon K]$. Since $v(\pi) > 0$ it follows $v(1+\pi c) = 0$ for every $c \in \mathcal{O}[\Upsilon K]$. Consequently if $a = \frac{b}{1+\pi c}$ is a typical element of R then $v(a) = v(b) \geq 0$. Hence v lies above R. The center of v on R is a prime ideal which contains π and is, therefore, a maximal ideal by what has been shown before.

q.e.d.

For any non-empty set S of valuations of K we denote by $\mathcal{O}_S$ the intersection of their valuation rings:

$$\mathcal{O}_S = \bigcap_{v \in S} \mathcal{O}_v \quad .$$

$\mathcal{O}_S$ is called the <u>holomorphy ring of</u> S <u>in</u> K. Every such holomorphy ring is integrally closed in K. Conversely, if R is any subring integrally closed in K then R is the holomorphy ring of a suitable set $S = S_R$ of valuations of K. Indeed, S_R is the set of those valuations of K which lie above R and are centered at some maximal ideal of R. In general, if $R \subset K$ is not necessarily integrally closed in K then the holomorphy ring of S_R equals the integral closure R^* of R in K ([Z-S],II, Chap VI,§4,Thm.6,p.15). Hence from Theorem 6.8 we obtain the following corollary:

COROLLARY 6.9 *In the same situation as in Theorem 6.8 let* $\mathcal{T}_{e,f}$ *denote the set of those p-valuations of* K *over* k *which are of relative type* (e,f) *over* k. *Then the holomorphy ring of* $\mathcal{T}_{e,f}$ *equals the integral closure* R^* *of the* γ-*Kochen ring* R *of* $K|k$.

There arises the question as to whether the γ-Kochen ring R itself is the holomorphy ring of $\mathcal{T}_{e,f}$; this means that R is integrally closed. In general this is not the case. However this is so if $e = 1$. To see this we need the following lemmas. In these lemmas we consider the same notations without further explanation. As far as possible we shall carry our discussion for arbitrary relative initial index ≥ 1.

LEMMA 6.10 *Let* P *be a maximal ideal of the* γ-*Kochen ring* R *of* $K|k$, *and let* v *be a valuation of* K *lying above* R *and centered at* P. *Hence by Theorem 6.8,* v *is a p-valuation over* k, *of relative type* (e,f); *in particular its relative initial index* e_v' *over* k *satisfies* $e_v' \leq e$. *If*

$$e_v' = e$$

then v *is the only valuation of* K *which lies over* R *and is centered at* P. *Thus* v *is then uniquely determined by its center* P *on* R.

In this context we do not distinguish between equivalent valuations, as it is evident from the formulation of the lemma. Accordingly if two valuations of K are considered different then they have different valuation rings.

Proof: Let w be another valuation of K lying above R and centered at some maximal idéal Q of R. Again, w is a p-valuation over k, of relative type (e,f). We assume $v \neq w$, i.e. $\mathcal{O}_v \neq \mathcal{O}_w$, and we claim that $P \neq Q$.

If $\mathcal{O}_w \subset \mathcal{O}_v$ then the image of $\mathcal{O}_w$ in the residue field $\bar{K}_v$ is a proper valuation ring of $\bar{K}_v$. But since $\bar{K}_v$ is finite, it does not admit any proper valuation ring. This shows that, indeed, $\mathcal{O}_w \not\subset \mathcal{O}_v$. Similarly $\mathcal{O}_v \not\subset \mathcal{O}_w$. Thus there exist $y,z \in K$ such that

$$v(y) \geq 0 \quad , \quad w(y) < 0$$

$$v(z) < 0 \quad , \quad w(z) \geq 0 .$$

If $v(y) > 0$ then we replace y by $y-1$ and hence we may assume $v(y) = 0$. Similarly we may assume $w(z) = 0$. Putting $x = \frac{y}{z}$ we obtain

$$v(x) > 0 \quad , \quad w(x) < 0 .$$

Here we may assume that x is a prime element for v. For if x were not a prime element, let Π be a prime element for v, and write $x = \Pi + (x - \Pi)$. Both Π and $x - \Pi$ are prime elements for v. Since $w(x) < 0$ we have $w(\Pi) < 0$ or $w(x - \Pi) < 0$. Hence after replacing x by Π or by $x - \Pi$ we may assume, indeed, that x is a prime element for v. We identify $v(x) = 1$ in the value group vK. Similarly in the value group wK, the value of a prime element for w is identified with 1. Hence $w(x) < 0$ implies $w(x) \leq -1$. Thus we have

$$v(x) = 1 \quad , \quad w(x) \leq -1 .$$

Now we refer to cases (i) resp. (ii) of the value list for $v(\beta x)$ as listed in Lemma 6.3. We conclude

$$v(\beta x) = 1 \quad , \quad w(\beta x) = -q\cdot w(x) \geq q .$$

Since $\gamma(x) = \frac{\beta(x)^e}{\pi}$ it follows

$$v(\gamma x) = e - v(\pi) \quad , \quad w(\gamma x) \geq e\cdot q - w(\pi) .$$

Observe that $v(\pi) = e_v'$, the relative initial index over k. By the hypothesis of the lemma $e_v' = e$. Since we have $w(\pi) = e_w' \leq e$, we obtain:

$$v(\gamma x) = 0 \quad , \quad w(\gamma x) \geq e\cdot q - e > 0 .$$

This shows that the element $\gamma(x) \in R$ is not contained in the center P of v, but it is contained in the center Q of w. Hence $P \neq Q$ as contended.

q.e.d.

Remark 6.11: *If $e = 1$ then the additional hypothesis $e_v' = e$ of Lemma 6.10 is automatically satisfied. Hence in this case, Lemma 6.10 yields a 1-1 correspondence between the maximal ideals of R and the p-valuations of $K|k$ of relative type $(1,f)$.*

However, if $e > 1$ and $e_v' < e$, then the conclusion of Lemma 6.10 does not hold in general. To see this let $\gamma'(X)$ denote the Kochen operator over k of type (e_v', f) ; then

$$\gamma(X) = \gamma'(X) \cdot \beta(X)^{e-e_v'}$$

where $\beta(X)$ is as in Lemma 6.3 . If we apply Lemma 6.2 to $\gamma'(X)$ then we obtain $v(\gamma'a) \geq 0$ for every $a \in K$. On the other hand $v(\beta a) > 0$ from Lemma 6.3 . We conclude

$$v(\gamma a) > 0$$

for every $a \in K$. We see that the center P_v of v on R contains the R-ideal $\langle\gamma K\rangle$ which is generated by the set γK . Since P_v contains π , we see that P_v contains the ideal $A = R\cdot\pi + \langle\gamma K\rangle$. This latter ideal A is <u>maximal</u> in R . To see this let us recall the definition of the Kochen ring:

$$R = \frac{\mathfrak{o}[\gamma K]}{1 + \pi\mathfrak{o}[\gamma K]} \quad .$$

This shows that every element $a \in R$ is congruent, modulo A, to some element $b \in \mathfrak{o}$. Hence $R/A = \mathfrak{o}/A \cap \mathfrak{o}$; since $A \cap \mathfrak{o}$ contains π we conclude $R/A = \bar{k}$, the residue field of $\mathfrak{o}$ modulo its maximal ideal. (Note that A does not contain 1 since $A \subset P_v$.) Consequently, A being maximal we see that $P_v = A$, which is an ideal independent of v . Hence if there is at least one other p-valuation $w \neq v$ of $K|k$, also of relative type (e,f) and also with $e_w' < e$, then $P_v = P_w = A$, showing that the conclusion of Lemma 6.10 does not hold in this case.

Now we turn to the discussion of the residue field.

LEMMA 6.12 *Let* P *be a maximal ideal of the* γ*-Kochen ring* R *of* $K|k$ *, and let* v *be a valuation of* K *lying above* R *and centered at* P *. Then* $R/P \subset \bar{K}_v$ *. By Theorem 6.8 ,* v *is a p-valuation over* k *, of relative type* (e,f) *over* k *; in particular* $e_v' \leq e$ *. If*

$$e_v' = e$$

then R/P *contains the* e*-th powers of all elements of* $\bar{K}_v$ *. In particular, if* $e = 1$ *then* $R/P = \bar{K}_v$ *.*

Proof: Let $x \in K$ be a prime element for v . As in the proof of Lemma 6.10 we see that $v(\gamma x) = 0$. Since $\gamma x \in R$ if follows that its residue class $\overline{\gamma x}$ is contained in R/P , and $\overline{\gamma x} \neq 0$.

Let $\bar{a}$ be an arbitrary non-zero element in $\bar{K}_v$, and let $a \in K$ be a foreimage of $\bar{a}$. We put $y = ax$; then y is also a prime element for v and therefore $\overline{\gamma y} \in R/P$. Let us compare $\overline{\gamma y}$ with $\overline{\gamma x}$. We have

$$\gamma(x) = \frac{\beta(x)^e}{\pi} = \frac{x^e}{\pi} \cdot \beta_o(x)^e$$

where

$$\beta_o(x) = \frac{\beta(x)}{x} = \frac{x^{q-1}-1}{(x^q-x)^2-1} .$$

We note that $\overline{\beta_o(x)} = 1$. Similarly $\overline{\beta_o(y)} = 1$. Hence we compute:

$$\frac{\gamma(y)}{\gamma(x)} = \left(\frac{y}{x}\right)^e \cdot \left(\frac{\beta_o(y)}{\beta_o(x)}\right)^e = a^e \cdot \left(\frac{\beta_o(y)}{\beta_o(x)}\right)^e$$

and therefore

$$\frac{\overline{\Upsilon(y)}}{\overline{\Upsilon(x)}} = \bar{a}^e \quad .$$

Consequently $\bar{a}^e \in R/P$, as contended.

q.e.d.

LEMMA 6.13 Let P be a maximal ideal of the Υ-Kochen ring R of $K|k$, and let v be a valuation of K lying above R and centered at P . Thus the quotient ring $\mathrm{Quot}_P(R)$ is contained in the valuation ring $\mathcal{O}_v$. If $e = 1$ then $\mathrm{Quot}_P(R) = \mathcal{O}_v$.

Proof: For brevity we write $R' = \mathrm{Quot}_P(R)$. Then R' is a local ring; its only maximal ideal is $P' = P \cdot R'$. The residue class field is $R'/P' = R/P$. The valuation v lies above R' and is centered at P' on R'. Since $e = 1$ we conclude from Lemma 6.12 that $R'/P' = \bar{K}_v = \mathcal{O}_v/\mathfrak{m}_v$ (where $\mathfrak{m}_v$ denotes the maximal ideal of the valuation ring $\mathcal{O}_v$). This can also be written in the form:

$$\mathcal{O}_v = R' + \mathfrak{m}_v \; .$$

Consequently, in order to show that $R' = \mathcal{O}_v$ it suffices to show that $\mathfrak{m}_v \subset R'$. Let $x \in \mathfrak{m}_v$. We put

$$R'' = R'[x]$$

and have to show that $R'' = R'$.

If w is any valuation of K lying above R' and centered at P' on R' then w lies above R and is centered on R at $P' \cap R = P$. Consequently, since $e = 1$ we infer from Lemma 6.10

that $w = v$. (See also Remark 6.11.) In other words: v is the <u>only</u> valuation of K which lies above R' and is centered at the maximal ideal of R'. This implies:

$\mathcal{O}_v$ <u>is the integral closure of</u> R' <u>in</u> K .

In particular it follows that x is integral over R' , hence $R'' = R'[x]$ is a finite R'-module. Since R' is local with maximal ideal P' we may apply <u>Nakayama's Lemma</u> and conclude: in order to show that $R'' = R'$ it suffices to show that $R'' = R' + P'\cdot R''$. Now we have $R'' = R'[x] = R' + x\cdot R''$; hence we have to show that

$$x \in P'\cdot R'' .$$

Let $P'' = \mathfrak{M}_v \cap R''$ be the center of v on R''. Then P'' is a prime ideal of R'' which contains P' . We claim that P'' is the <u>only</u> prime ideal of R'' containing P' . To see this, let Q'' be another prime ideal of R'' containing P' . There exists a valuation w of K lying above R'' and centered at Q'' on R''. Then w lies also above R' and is centered on R' at the prime ideal $Q'' \cap R' \supset P'$. Since P' is maximal in R' it follows that $Q'' \cap R' = P'$, hence w is centered at P' on R' . We conclude $w = v$ since we know that v is the only valuation of K which lies above R' and is centered at P' on R' . Since $w = v$ we have indeed $Q'' = P''$.

Now since P'' is the only prime ideal of R'' containing P', it follows that every element of P'' is nilpotent modulo $P'\cdot R''$. In particular we have

$$x^n \in P'\cdot R''$$

for some natural number $n > 0$. Our claim is that this holds for $n = 1$. This is shown as follows.

The relation $x^n \in P' \cdot R'' = P'[x]$ says that x^n can be represented in the form

$$x^n = h(x)$$

where $h(X)$ is a polynomial with coefficients in P'. If we put $g(X) = X^n - h(X)$ then $g(X) \in R'[X]$ and:

$$g(x) = 0 \quad , \quad \bar{g}(X) = X^n .$$

As usual, $\bar{g}(X)$ denotes the polynomial in $\bar{K}_v[X]$ obtained from $g(X)$ by applying the residue map $\mathcal{O}_v \to \bar{K}_v$ to its coefficients.

Now let N denote the ideal of <u>relations</u> for x over R'; it consists of all polynomials $f(X) \in R'[X]$ with $f(x) = 0$. Let $\bar{N}$ be the image of N with respect to the natural map $R'[X] \to \bar{K}_v[X]$. Note that this map is surjective since $R'/P' = \bar{K}_v$, as observed earlier. Hence $\bar{N}$ is an ideal of $\bar{K}_v[X]$. The above relations for $g(X)$ say that $g(X) \in N$ and $X^n \in \bar{N}$.

Another polynomial in $\bar{N}$ can be constructed as follows. By definition of the γ-Kochen ring R we have $\gamma(x) \in R \subset R'$. Let us put $\gamma(x) = a$. Then since $e = 1$:

$$\pi a = \beta(x) = \frac{x^q - x}{(x^q-x)^2 - 1} .$$

Thus x is a zero of the polynomial

$$\varphi(X) = \pi a \cdot (X^q - X)^2 - (X^q - X) - \pi a$$

which has coefficients in R' . Hence $\varphi(X) \in N$ and

$$\bar{\varphi}(X) = -(X^q - X) \in \bar{N} .$$

We have shown that the ideal $\bar{N} \subset \bar{K}_v[X]$ contains the two polynomials X^n and $X^q - X$. Hence $\bar{N}$ contains their greatest common divisor, which is X . The relation $X \in \bar{N}$ says that there exists a polynomial $f(X) \in N$ of the form

$$f(X) = p_o + X + p_2X^2 + \ldots + p_rX^r$$

with $p_i \in P'$. (Originally we should admit a coefficient c at X such that $\bar{c} = 1$; but then c is a unit in R' and hence, after dividing by c we may assume that X appears in $f(X)$ with coefficient 1 .) Since $f(x) = 0$ we obtain

$$x = -p_o - p_2x^2 - \ldots - p_rx^r \in P'[x] = P' \cdot R'' .$$

This proves our assertion.

q.e.d.

We are now ready to state the Spectral Structure Theorem for the γ-Kochen ring. Let us recall the situation which we have studied in this section.

k is a given p-valued field, with prime element π , residue order q_k and valuation ring $\mathfrak{o}$. Let e and f be natural numbers and consider the π-adic Kochen operator of type (e,f):

$$\gamma(X) = \frac{\beta(X)}{\pi}^e$$

where

$$\beta(X) = \frac{X^q - X}{(X^q - X)^2 - 1} \quad , \quad q = q_k^f \quad .$$

Let K be a field extension of k , and let R be the γ-Kochen ring of K :

$$R = \frac{\mathcal{O}[\gamma K]}{1 + \pi \cdot \mathcal{O}[\gamma K]} \quad .$$

It is assumed that $R \neq K$. This is equivalent to the fact that K admits p-valuations v over k which are of relative type (e,f) over k . (This means that the relative initial index e_v' over k is $\leq e$, and that the relative residue degree f_v' over k divides f .) Let $\mathcal{T}_{e,f}$ denote the set of those valuations v .

THEOREM 6.14 (Spectral Structure Theorem for the Kochen ring) In the situation described above, assume in addition that $e = 1$. Then the γ-Kochen ring R of K is integrally closed, and R is the holomorphy ring of the set $\mathcal{T}_{1,f}$ of valuations of K . There is a bijective correspondence between the maximal ideal spectrum of R and the set $\mathcal{T}_{1,f}$. If the maximal ideal P of R and the valuation $v \in \mathcal{T}_{1,f}$ correspond to each other then P is the center of v on R , and $\mathcal{O}_v$ is the quotient ring of R with respect to P :

$$P = \mathfrak{m}_v \cap R \quad , \quad \mathcal{O}_v = \mathrm{Quot}_P(R) \quad .$$

The proof is obtained by collecting the results of the preceding lemmas and theorems. Let $v \in \mathcal{T}_{1,f}$. By Theorem 6.8 , v lies above R and is centered at some maximal ideal P of R .

By Lemma 6.13 , $\mathcal{O}_v = \mathrm{Quot}_P(R)$. Hence v is uniquely determined by its center P . Again by Theorem 6.8 , every maximal ideal P of R appears as the center of some $v \in \mathcal{T}_{1,f}$. Finally, since an integral domain is the intersection of its quotient rings with respect to its maximal ideals, it follows $R = \bigcap \mathcal{O}_v$ where v ranges over $\mathcal{T}_{1,f}$. Hence R is the holomorphy ring of $\mathcal{T}_{1,f}$ and R is integrally closed in K .

Problem: Does there exist an operator $\tilde{\gamma}_{e,f}(X)$ such that Theorem 6.14 holds for the corresponding Kochen ring $R_{\tilde{\gamma}}$, for arbitrary e and f ?

In 6.3 we will continue to study the ideal theory of the Kochen ring. Now, let us turn for a moment to the case where K is p-adically closed or, more general, where K is henselian with respect to some p-valuation v .

THEOREM 6.15 In the situation as described before Theorem 6.14, assume in addition that K is henselian with respect to some p-valuation v of relative type (e,f) over k . Then v is the only p-valuation on K whatsoever.

If $e = 1$ then $\mathcal{O}_v = \gamma K$ where γ is the π-adic Kochen operator over k of type $(1,f)$. In particular, if we let $K = k$, then $\mathcal{O}_v = \gamma K$ where γ is of type $(1,1)$.

Proof: Let w be some p-valuation of K , different from v . The compositum $\mathcal{O}$ of the valuation rings $\mathcal{O}_v$ and $\mathcal{O}_w$ is again a valuation ring of K. Since $\mathcal{O}_v$ and $\mathcal{O}_w$ both have finite residue field, neither of them can be equal to $\mathcal{O}$ (cf. the argument in the proof of Lemma 6.10). Thus $\mathcal{O}$, as a proper extension of $\mathcal{O}_v$, must contain the coarse valuation ring $\mathcal{O}_v$. In particular, $\mathcal{O}$ has residue characteristic zero. Passing to the residue field of $\mathcal{O}$, the images of $\mathcal{O}_v$ and $\mathcal{O}_w$ are then independent p-valuation rings (the image of $\mathcal{O}_v$ being still henselian). Thus we may assume from the beginning that $\mathcal{O}_v$ and $\mathcal{O}_w$ have compositum K. Therefore $\mathfrak{M}_v \cap \mathcal{O}_w \not\subset \mathfrak{M}_w$ and hence we can find some unit a of $\mathcal{O}_w$ belonging also to $\mathfrak{M}_v$. We consider the polynomial

$$f(X) = X^{q'} - X + a ,$$

where we have chosen q' as a power of p such that all elements x of the residue fields of $\mathcal{O}_v$ and $\mathcal{O}_w$ satisfy $x^{q'} - x = 0$. Since $\mathcal{O}_v$ is henselian, the polynomial $f(X)$ has a zero in K, say b. (Observe that 1 is a simple zero of $f(X)$ in the residue field of $\mathcal{O}_v$.) Passing now to the residue field of $\mathcal{O}_w$ we obtain the contradiction

$$\bar{a} = \bar{b}^{q'} - \bar{b} + \bar{a} = \bar{0} .$$

Now assume in addition that $e = 1$. For every $a \in \mathcal{O}_v$ the polynomial

$$g(X) = a\pi[(X^q - X)^2 - 1] - (X^q - X)$$

admits 1 as a simple zero in the residue field of $\mathcal{O}_v$. By Hensel's Lemma $g(X)$ has a zero in K, say b. Therefore we obtain

$$a = \frac{1}{\pi} \cdot \frac{b^q - b}{(b^q - b)^2 - 1} = \gamma(b) .$$

From the obvious inclusion $\Upsilon K \subset \mathcal{O}_V$ we thus conclude $\Upsilon K = \mathcal{O}_V$.

6.3 The Principle Ideal Theorem

Recall that an integral domain is called Prüfer ring if every finitely generated fractional ideal is invertible. It is well known that Prüfer rings can be characterized by the fact that for every maximal ideal, the corresponding quotient ring is a valuation ring. Hence we obtain from Theorem 6.14

COROLLARY 6.16 *In the situation of Theorem 6.14 the Υ-Kochen ring R is a Prüfer ring.*

We have just seen that in certain cases, the Υ-Kochen ring R is a Prüfer ring. We shall now show that in these cases R has the following stronger property: *Every finitely generated ideal of R is pincipal*. Such rings are called *Bezout rings*.

It will turn out that the Bezout property holds always for the holomorphy rings of the sets $\mathcal{T}_{e,f}$ (as defined in the preceding section), even if $e > 1$ and hence the holomorphy ring may perhaps not be equal to the Kochen ring. In fact, there is the following general theorem.

THEOREM 6.17 (<u>Principal Ideal Theorem</u>) <u>Let</u> K <u>be any field and let</u> S <u>be a non-empty set of valuations of</u> K <u>with the property that the residue fields</u> $\bar{K}_v$ <u>for</u> $v \in S$ <u>are finite, and that their orders</u> $q_v = |\bar{K}_v|$ <u>are bounded. Then the holomorphy ring</u> $\mathcal{O}_S \subset K$ <u>is a Bezout ring with</u> K <u>as its field of quotients</u>.

<u>Proof</u>: By assumption there is an integer q such that $q_v \leq q$ for all $v \in S$. After enlarging q if necessary we may assume that $q_v - 1 \mid q - 1$ for all $v \in S$. Then the polynomial $X^q - X$ vanishes on all the residue fields $\bar{K}_v$ with $v \in S$. For if $0 \neq \bar{a} \in \bar{K}_v$ then $\bar{a}$ is an $(q_v - 1)$-th root of unity; since $q_v - 1 \mid q - 1$ it follows $\bar{a}^{q-1} = 1$ and thus $\bar{a}^q = \bar{a}$. We conclude that neither of the two polynomials

$$f_1(X) = 1 + (X^q - X)$$

$$f_2(X) = 1 + X \cdot (X^q - X)$$

has a root in $\bar{K}_v$, for all $v \in S$. In the following we use the symbol $f(X)$ to denote either $f_1(X)$ or $f_2(X)$; this notation serves to avoid duplication of the arguments. Let n denote the degree of $f(X)$; thus we have $n = q$ if $f(X) = f_1(X)$ and $n = q + 1$ if $f(X) = f_2(X)$.

Let $\mathcal{O} = \mathcal{O}_S$ denote the holomorphy ring of the given set S of valuations of K. Our contention is that every finitely generated ideal is principal. Clearly it suffices to show: every $\mathcal{O}$-ideal <u>generated by two elements</u> is principal. Let $0 \neq a, b \in K$

and consider the $\mathcal{O}$-module $A = \mathcal{O}a + \mathcal{O}b$. We want to find $c \in K$ such that $A = \mathcal{O}c$. After multiplying with b^{-1} we may assume without loss that $b = 1$. Thus we have now $A = \mathcal{O} + \mathcal{O}a$.

Consider the n-th power

$$A^n = \mathcal{O} + \mathcal{O}a + \mathcal{O}a^2 + \ldots + \mathcal{O}a^n .$$

We claim that A^n is principal, and is generated by $f(a)$. Let us write

$$f(X) = c_o + c_1 X + \ldots + c_{n-1} X^{n-1} + X^n ,$$

with $c_i \in \mathbb{Z}$. Then the element

$$f(a) = c_o + c_1 a + \ldots + c_{n-1} a^{n-1} + a^n$$

is contained in A^n. In order to show that $f(a)$ generates A^n we have to show that

$$a^i \in \mathcal{O}f(a) \qquad (0 \leq i \leq n) .$$

Since $\mathcal{O} = \bigcap_{v \in S} \mathcal{O}_v$ this is equivalent to

$$v(a^i) \geq v(f(a)) \qquad (0 \leq i \leq n) ,$$

for each $v \in S$. In order to verify these relations we distinguish two cases, according to whether $v(a) \geq 0$ or $v(a) < 0$.

<u>Case 1</u>: $v(a) \geq 0$. Then $v(a^i) \geq 0$ if $i \geq 0$ and thus we have to show that

$$v(f(a)) = 0 .$$

Indeed, if $v(f(a)) > 0$ then $f(\bar{a}) = 0$ where $\bar{a} \in \bar{K}_v$ denotes the residue class of a. But this contradicts the fact that $f(X)$ does not admit any root in $\bar{K}_v$ for $v \in S$.

Case 2: $v(a) < 0$. Then $v(a^i) \geq v(a^n)$ if $i \leq n$ and thus we want to show that $v(a^n) = v(f(a))$ which is to say that

$$0 = v\left(\frac{f(a)}{a^n}\right) \quad .$$

Indeed, we have

$$\frac{f(a)}{a^n} = c_o a^{-n} + c_1 a^{-(n-1)} + \ldots + c_{n-1} a^{-1} + 1$$

and $v(c_i a^{-(n-i)}) \geq v(a^{-(n-i)}) > 0$ if $i < n$.

We have now proved that $A^n = \mathcal{O} f(a)$. This shows in particular that $f(a) \neq 0$; hence

$$A^n \cdot f(a)^{-1} = \mathcal{O} \quad .$$

The above arguments hold for $f(X) = f_1(X)$ (with $n = q$) and $f(X) = f_2(X)$ (with $n = q+1$). We conclude:

$$A^q = \mathcal{O} f_1(a)$$

$$A^{q+1} \cdot f_2(a)^{-1} = \mathcal{O}$$

hence

$$A \cdot f_1(a) \cdot f_2(a)^{-1} = A \cdot A^q \cdot f_2(a)^{-1} = \mathcal{O}$$

$$A = \mathcal{O} \cdot f_2(a) \cdot f_1(a)^{-1} \quad .$$

Thus the $\mathcal{O}$-module $A = \mathcal{O} + \mathcal{O} a$ is principal, generated by the element $z = f_2(a) \cdot f_1(a)^{-1}$.

Since $1 \in \mathcal{O}z$ it follows $z^{-1} \in \mathcal{O}$; since $a \in \mathcal{O}z$ it follows $az^{-1} \in \mathcal{O}$. We conclude that

$$a = \frac{az^{-1}}{z^{-1}}$$

is contained in the quotient field of $\mathcal{O}$. Here, a is an arbitrary non-zero element of K . Therefore:

$$K = \mathrm{Quot}(\mathcal{O}) .$$

Also, we see that $A = \mathcal{O} + \mathcal{O}a$ can now be regarded as a fractional $\mathcal{O}$-ideal (not only as an $\mathcal{O}$-module). And we have proved that every fractional $\mathcal{O}$-ideal generated by two elements is principal.

q.e.d.

We remark that the Principal Ideal Theorem can be shown to hold in a much more general situation, even with infinite residue fields. For details see [R].

Returning to the situation of Theorem 6.14 , we now know that in that situation the γ-Kochen ring R is a Bezout ring. It follows then that <u>every overring of the γ-Kochen ring in K is a Bezout ring too</u>. Indeed, every overring of a Bezout domain in its field of quotients is a Bezout ring.

§ 7. Function fields over p-adically closed fields

We will now deal with function fields $K|k$ in n variables, i.e. K is finitely generated over k of transcendence degree n, where n is a natural number. The field k will always be assumed to carry a p-valuation ring $\mathfrak{o}$ of p-rank d with prime element π. Mostly we will assume that k is p-adically closed. In that case we know from Theorem 6.15 that $\mathfrak{o}$ is the only p-valuation ring on k and $\mathfrak{o} = \gamma k$ where γ is the π-adic Kochen operator

$$\gamma(X) = \frac{1}{\pi} \cdot [(X^q - X) - (X^q - X)^{-1}]^{-1}$$

with $q = p^f$, f being the degree of $\bar{k}$ over its prime field. In particular, any p-valuation of K is an extension of $\mathfrak{o}$.

If the function field $K|k$ admits some rational place, Example 2.5 yields some p-valuation of K of p-rank d extending $\mathfrak{o}$. Hence K is formally p-adic of relative type $(1,1)$ over $\mathfrak{o}$. We will prove the converse under the assumption that k is p-adically closed, i.e. under this assumption we will prove the existence of rational places of $K|k$, if K is formally p-adic of relative type $(1,1)$. Therefore, in this section, on K we will only consider p-valuations of relative type $(1,1)$ over $\mathfrak{o}$, i.e. p-valuations of p-rank d extending $\mathfrak{o}$.

The γ-Kochen ring R_γ of K over k (of relative type $(1,1)$) is the subring of K generated by elements of the form

$$a = \frac{b}{1+\pi c} \qquad \text{with } b,c \in \mathcal{O}[\Upsilon K] \text{ and } \quad 1+\pi c \neq 0 \; .$$

By Theorem 6.14, it is equal to the intersection of all p-valuation rings $\mathcal{O}$ of K of relative type (1,1) extending $\mathcal{O}$. As an analogy to the 17th problem of Hilbert (see Section 1) we will prove that a polynomial $f(X_1,\ldots,X_n) \in k[X_1,\ldots,X_n]$ belongs to R_Υ if and only if it is <u>integral definite over</u> k , i.e. $f(a_1,\ldots,a_n) \in \mathcal{O}$ for all $a_1,\ldots,a_n \in k$.

7.1 Existence of rational places

We assume the reader to be familiar with the basic notions and facts about places. By a place of the function field $K|k$ in n variables we mean a place of K which is the identity on k . For any such place P , its residue field KP is an extension field of k . The transcendence degree m of KP over k will be called the <u>dimension</u> of P . We have $n = m$ if and only if P is trivial on K ; otherwise $m < n$. In general, KP need not be finitely generated over k . In case $n = 1$ it is well-known for non-trivial P that KP is finite algebraic over k and that the corresponding value group is isomorphic to $\mathbb{Z}$, i.e. discrete of rank one. More generally, in case $m = n-1$, KP is finitely generated and the corresponding value group is again $\mathbb{Z}$. Indeed, if $y_1,\ldots,y_m$ is a transcendence base of KP over k and $x_1,\ldots,x_m$ are preimages of $y_1,\ldots,y_m$, then P is trivial on $k(x_1,\ldots,x_m)$ and the result follows from the case $n = 1$ applied to the function field $K|k(x_1,\ldots,x_m)$.

A place P of the function field $K|k$ is called rational if $KP = k$. Thus in particular a rational place of $K|k$ is of dimension 0. The space of all rational places on $K|k$ is called the Riemann k-space of $K|k$ and is denoted by $S_k(K|k)$. In order to simplify the notation we shall use the symbol S instead of $S_k(K|k)$, and shall briefly speak of the "Riemann space".

The image of $x \in K$ with respect to a place P will be denoted by xP. If P ranges over S we then obtain a function $x: S \to k \cup \{\infty\}$. This function is continuous with respect to the p-adic topologies canonically defined on S and on $k \cup \{\infty\}$. Let $u = (u_1, \ldots, u_r)$ be a finite family of elements of K, and consider the set S_u of those places $P \in S$ which satisfy the integrality condition

$$u_j P \in \mathcal{O} \qquad (1 \leq j \leq r) .$$

The subsets S_u of S obtained in this way can be taken as basic open sets for the p-adic topology of S. Note that S_u is open and closed in S. If u is empty then $S_u = S$.

There is also the Zariski topology on S. Let $z = (z_1, \ldots, z_s)$ be a finite family of elements in K. Consider the set S^z of those places $P \in S$ which satisfy the holomorphy condition

$$z_i P \neq \infty \qquad (1 \leq i \leq s) .$$

The subsets S^z of S obtained in this way can be taken as basic open sets for the Zariski topology on S. If z is empty then $S^z = S$. The intersection of a basic Zariski-open set S^z with a basic p-adic open set S_u will be called a <u>basic subset</u> of S ; we write

$$S_u^z = S^z \cap S_u .$$

If S_u is non-empty, say $P \in S_u$, then the preimage of $\mathfrak{o}$ with respect to P is a p-valuation ring $\mathcal{O}$ on K of p-rank d such that $u_j \in \mathcal{O}$ for $1 \leq j \leq r$. If there is a p-valuation ring $\mathcal{O}$ extending $\mathfrak{o}$ with $u_j \in \mathcal{O}$ $(1 \leq j \leq r)$ and of p-rank d, then K will be called <u>formally p-adic over</u> $\mathfrak{o}[u]$.

LEMMA 7.1 <u>In the situation explained above,</u> K <u>is formally p-adic over</u> $\mathfrak{o}[u]$ <u>if and only if</u> $\pi^{-1} \notin \mathfrak{o}[u,\gamma K]$.

<u>Proof</u>: If $\mathcal{O}$ is a p-valuation on K of p-rank d extending $\mathfrak{o}[u]$, then $\mathcal{O}$ is of relative type (1,1) and thus by Lemma 6.2 contains $\mathfrak{o}[\gamma K]$. Therefore $\mathfrak{o}[u,\gamma K]$ is contained in $\mathcal{O}$ which implies $\pi^{-1} \notin \mathfrak{o}[u,\gamma K]$.

Conversely, if π is not a unit in $\mathfrak{o}[u,\gamma K]$, we conclude as in the proof of Theorem 6.4 from general valuation theory that there is a valuation ring of K containing $\mathfrak{o}[u,\gamma K]$ and centered over π . But then it also follows from 6.4 that $\mathcal{O}$ is a p-valuation of relative type (1,1). Hence $\mathcal{O}$ has p-rank d and contains $\mathfrak{o}[u]$.

q.e.d.

We now come to the main theorem of this subsection.

THEOREM 7.2 (<u>Place Existence Theorem</u>) <u>Let</u> $K|k$ <u>be a function field in</u> n <u>variables over the p-adically closed field</u> k . <u>Then</u>

<u>the basic subset</u> S_u^z <u>of the Riemann space</u> S <u>is non-empty if and only if</u> K <u>is formally p-adic over</u> $\mathfrak{o}[u]$.

<u>Proof</u>: By the observation before Lemma 7.1 it remains to prove the existence of some $P \in S_u^z$ under the assumption that K is formally p-adic over $\mathfrak{o}[u]$. This proof will proceed by induction on n . More precisely, we will show the following:

(1) if $n = 1$ then $S_u^z \neq \emptyset$;

(2) there exists a place P of $K|k$ of dimension $n-1$ such that $z_i P \neq \infty$, $u_j P \neq \infty$, and KP is formally p-adic over $\mathfrak{o}[uP]$.

Here we use uP as an abbreviation for the family $(u_1 P, \ldots, u_r P)$. It is clear how to proceed now by induction: let Q be a member of the basic subset S_{uP}^{zP} of the Riemann space of $KP|k$ which exists by the induction hypothesis. Then the composite place $P \circ Q$ belongs to S_u^z .

Let us begin with the proof of (1) . K may be considered as the field of quotients of $k[X,Y]/(f)$ where $f \in k[X,Y]$ is irreducible and monic as a polynomial in Y . Let us denote the residue classes of X and Y with respect to the ideal (f) by x and y respectively. Then $K = \mathrm{Quot}(k[x,y]) = k(x,y)$. The functions u_j $(1 \leq j \leq r)$ and z_i $(1 \leq i \leq s)$ can be represented as quotients

$$u_j = \frac{u_j'}{u_j''} \quad \text{and} \quad z_i = \frac{z_i'}{z_i''}$$

with $u_j', u_j'', z_i', z_i'' \in k[x,y]$.

By the assumption of the theorem, K is formally p-adic over

$\mathfrak{o}[u]$. Thus there is a p-valuation ring $\mathcal{O}$ on K of p-rank d containing $\mathfrak{o}[u]$. If L denotes some p-adic closure of K with respect to $\mathcal{O}$, $L|k$ is an extension of p-adically closed fields of p-rank d. Thus we are in the situation of Theorem 5.1. Now every elementary sentence about valued fields (with parameters from k) which holds in L also holds in k. We apply this property of $L|k$ to the following statement about L:

(+) there exist x and y such that

$$f(x,y) = 0\ ,\ f_Y(x,y) \neq 0\ ,\ z_i''(x,y) \neq 0 \quad (1 \le i \le s)$$

$$u_j''(x,y) \neq 0\ ,\ u_j(x,y) \in \mathcal{O} \quad (1 \le j \le r)$$

where f_Y denotes the derivative of f with respect to Y. Clearly, this statement can be expressed by a sentence of the language of valued fields which thus holds in L. (Note that it already holds in K.) Applying Theorem 5.1 to this sentence we also know it to hold in k. Thus there are elements $a,b \in k$ satisfying

(*) $$f(a,b) = 0\ ,\quad f_Y(a,b) \neq 0$$

$$z_i''(a,b) \neq 0 \qquad (1 \le i \le s)$$

$$u_j''(a,b) \neq 0\ ,\quad u_j(a,b) \in \mathfrak{o} \quad (1 \le j \le r)\ .$$

Consider now the field $k((t))$ of formal Laurent series in t with coefficients in k (cf. Example 2.3). The polynomial $f(a+t,Y)$ has b as a simple zero in the residue field of the canonical place of $k((t))$. Indeed, the residue field is k and the image of $f(a+t,Y)$ under the residue map is $f(a,Y)$. Since $k((t))$ is Henselian with respect to the canonical place, $f(a+t,Y)$ has a zero $h(t)$ in $k((t))$ which can be written like

$$h(t) = b + b_1 t + b_2 t^2 + \dots \qquad (b_\nu \in k) .$$

Obviously, the prescriptions $x \mapsto a+t$ and $y \mapsto h(t)$ yield a k-embedding of $K = k(x,y)$ into $k((t))$. The canonical place of $k((t))$ now induces a rational place P of $K|k$ such that

$$xP = a \quad \text{and} \quad yP = b \quad .$$

From the properties of a,b we therefore obtain

$$z_i P \neq \infty \quad \text{and} \quad u_j P \in \mathfrak{o} .$$

This finishes the proof of (1).

The proof of (2) will be just an application of (1) to a suitably chosen function field. Let $x_1,\dots,x_n$ be a transcendence base of K over k and let $K = k(x_1,\dots,x_n,y)$. By the assumption of the theorem there is a p-valuation $\mathcal{O}$ on K of p-rank d containing $\mathfrak{o}[u]$. Let k_1 be the relative algebraic closure of $k(x_1,\dots,x_{n-1})$ inside some p-adic closure of K with respect to $\mathcal{O}$. Denote by $\mathfrak{o}_1$ the induced p-valuation on k_1 . Then by Theorem 3.4 , k_1 is p-adically closed with respect to $\mathfrak{o}_1$. Moreover, the field $k_1(x_n,y)$ is formally p-adic over $\mathfrak{o}_1[u]$. Thus we can apply (1) to the extension $k_1(x_n,y)|k_1$ in order to obtain a rational place P_1 of this extension such that

$$z_i P_1 \neq \infty \quad \text{and} \quad u_j P_1 \in \mathfrak{o}_1 \quad .$$

Finally, the restriction P of P_1 to the field K then is a place of $K|k$ of dimension $n-1$ such that $z_i P \neq \infty$, $u_j P \neq \infty$, and KP is formally p-adic over $\mathcal{O}[uP]$.

q.e.d.

In the above proof of (1) we have applied Theorem 5.1 to the sentence formalizing (+) . Because of the simplicity of (+) , however, only Theorem 4.3 is needed under certain special assumptions. We now try to figure those out. At the same time we give a self-contained proof of (1), and thus of Theorem 7.2 .

In the situation of (1), $K|k$ is a function field in one variable. k is p-adically closed with $\mathcal{O}$ as its unique p-valuation ring. K admits a p-valuation ring $\mathcal{O}$ of p-rank d, containing $\mathcal{O}[u]$. Using the notations from the above proof, we are looking for some $a,b \in k$ satisfying (*) . We will prove that such elements exist in k under the additional assumption that k _is countable_ .

Clearly, this is no restriction since we may replace k by the relative algebraic closure k_o of the subfield generated by the coefficients of the functions f, z_i', z_i'', u_j', u_j'' . This field k_o is again p-adically closed of p-rank d by Theorem 3.4 . Finally K is replaced by $k_o(x,y)$. If we can find elements $a,b \in k_o$ with the above properties, we are clearly done. Thus we can assume k to be countable.

PROPOSITION 7.3 _Let_ $K|k$ _satisfy the above mentioned conditions. Then_ K _embeds (over_ k_) into_ k^* _as a valued field where_ k^* _is the field constructed in Example_ 2.6 _with p-valuation ring_ $\mathcal{O}^*$ _of p-rank_ d.

Using the notations of Example 2.6 and Proposition 7.3 we can then argue as follows. Let

$$(a^{(n)})_{n\in\mathbb{N}} + M \quad \text{and} \quad (b^{(n)})_{n\in\mathbb{N}} + M$$

be the images of x and y under such an embedding of $K = k(x,y)$ into k^*. From the properties of x and y listed in (+) we see that the set of $n \in \mathbb{N}$ such that

$$f(a^{(n)},b^{(n)}) = 0\ , \quad f_Y(a^{(n)},b^{(n)}) \neq 0\ ,$$
$$u_i''(a^{(n)},b^{(n)}) \neq 0,\ u_i(a^{(n)},b^{(n)}) \in \mathcal{O},\ z_j''(a^{(n)},b^{(n)}) \neq 0$$

belongs to the ultra-filter D on $\mathbb{N}$. Since D is non-empty there is some $n_o \in \mathbb{N}$ with the above property. Now $a = a^{(n_o)}$ and $b = b^{(n_o)}$ are the elements of k we are looking for.

Proof: (Proposition 7.3) We will make use of the proof of Theorem 4.3 under the special situation that the base field k is p-adically closed and countable (in 4.3 denoted by K), the field K to be embedded is finitely generated of transcendence degree 1 over k (in 4.3 denoted by L), and the receiving field k^* is $\aleph_1$-saturated and p-adically closed (in 4.3 denoted by L'). It remains to prove these two properties of k^*. By Remark 4.12 it suffices to know that k^* is $\aleph_1$-saturated and Henselian in order to obtain the desired embedding.

Let us first show that k^* is Henselian. By the remark on page 22 we have to show that each polynomial

$$g(X) = X^{m+1} + X^m + \pi a_{m-1}X^{m-1} + \ldots + \pi a_o$$

with $a_i \in \mathcal{O}^*$ has a zero in k^*. Let

$$a_i = (a_i^{(n)})_{n\in\mathbb{N}} + M \in \mathcal{O}^*$$

Then the set $A = \{n | a_i^{(n)} \in \mathcal{O} \text{ for } 0 \leq i \leq m-1\}$ belongs to the ultra filter D. Thus for each $n \in A$, the polynomial

$$g^{(n)}(X) = X^{m+1} + X^m + \pi a_{m-1}^{(n)} X^{m-1} + \ldots + \pi a_o^{(n)}$$

has a zero in k, say $x^{(n)}$. Let $x^{(n)} = 0$ for $n \notin A$. Then the set $\{n | g^{(n)}(x^{(n)}) = 0\}$ contains A and thus also belongs to D. Therefore $x = (x^{(n)})_{n\in\mathbb{N}} + M$ is a zero of g in k^*.

In model theory it is well-known that k^* is $\aleph_1$-saturated whenever D is a non-principal ultra-filter on $\mathbb{N}$. However, in order to give a self-contained proof of the Place Existence Theorem, we will not make use of this fact. Rather we will single out from the proof of Theorem 4.3 what is really used from the saturation property and give an 'ad hoc' proof for this. Since K is already finitely generated of transcendence degree one over k, saturatedness enters into the proof of Theorem 4.3 only through Propositions 4.10 A and B. To be more precise it enters on pages 75 and 81.

On page 75 we have to find some $y \in k^*$ satisfying

$$v(x-a) = v(y-a)$$

for all $a \in k$ under the assumption that for any finite number of elements $a_1,\ldots,a_n \in k$ there is some $y \in k$ with

$$v(x-a_i) = v(y-a_i) \qquad (1 \leq i \leq n).$$

(Here and in the following v always denotes the p-valuation under consideration, no matter whether we are in K, k, or k^*.) Now let $(a_m)_{m\in\mathbb{N}}$ be an enumeration of all elements of k, and $b_m \in k$ such that $v(b_m) = v(x-a_m)$. Choose $y^{(n)} \in k$ such that

$$v(b_m) = v(y^{(n)} - a_m) \quad \text{for all} \quad m \leq n .$$

The element

$$y = (y^{(n)})_{n\in\mathbb{N}} + M$$

from k^* is the one we are looking for. Indeed, for a given $m \in \mathbb{N}$

$$A_m = \{n \mid v(b_m) = v(y^{(n)} - a_m)\} \supset \{n \mid n \geq m\} .$$

Since the ultra-filter D is non-principal, the cofinite set $\{n \mid n \geq m\}$ and hence also A_m belong to D. Thus for every $m \in \mathbb{N}$ we have

$$v(b_m) = v(y - a_m) .$$

This clearly follows from the above observation, since

$$A_m = \left\{n \;\middle|\; \frac{y^{(n)} - a_m}{b_m} \in \mathcal{O} \text{ and } \frac{b_m}{y^{(n)} - a_m} \in \mathcal{O}\right\}$$

and therefore

$$\frac{y - a_m}{b_m} \in \mathcal{O}^* \qquad \text{and} \qquad \frac{b_m}{y - a_m} \in \mathcal{O}^* .$$

On page 81 the argument is similar. There we have to find some $y \in k^*$ such that

$$\operatorname{sgn}(\xi - \alpha) = \operatorname{sgn}(vy - \alpha)$$

for all $\alpha \in \mathbb{Q} \otimes vk$ under the assumption that for every finite set $\alpha_1,\ldots,\alpha_n$ such a y already exists in k. (Here we use an observation pointed out in Remark 4.12.) Let $(\alpha_m)_{m\in\mathbb{N}}$ be an enumeration of all elements of $\mathbb{Q} \otimes vk$. Choose $y^{(n)} \in k$ such that

$$\operatorname{sgn}(\xi - \alpha_m) = \operatorname{sgn}(vy^{(n)} - \alpha_m) \quad \text{for all} \quad m \leq n .$$

As above, we conclude for fixed $m \in \mathbb{N}$ that

$$B_m = \{n \mid \operatorname{sgn}(\xi - \alpha_m) = \operatorname{sgn}(vy^{(n)} - \alpha_m)\} \in D .$$

Thus for every $m \in \mathbb{N}$ the element

$$y = (y^{(n)})_{n\in\mathbb{N}} + M$$

of k^* satisfies $\operatorname{sgn}(\xi - \alpha_m) = \operatorname{sgn}(vy - \alpha_m)$. To see this, one has to discuss three cases depending on the value of $\operatorname{sgn}(\xi - \alpha_m)$. Let us deal e.g. with the case $\operatorname{sgn}(\xi - \alpha_m) = 1$. Let us also assume that

$$\alpha_m = \frac{va}{r} \quad \text{for some} \quad a \in k, \ r \in \mathbb{N} .$$

Then

$$B_m = \{n \mid vy^{(n)} > \frac{va}{r}\} = \{n \mid \frac{(y^{(n)})^r}{\pi a} \in \mathcal{O}\} \in D .$$

Hence

$$\frac{y^r}{\pi a} \in \mathcal{O}^*$$

or equivalently $vy > \frac{va}{r}$ which implies

$$\operatorname{sgn}(\xi - \alpha_m) = \operatorname{sgn}(vy - \alpha_m) .$$

The other cases are treated similarly. q.e.d.

7.2 The holomorphy ring of a function field

We adopt the notations of Subsection 7.1 . An element $x \in K$ is called holomorphic at $P \in S$ if $xP \neq \infty$, i.e. if x is contained in the valuation ring $\mathcal{O}_P$ of K belonging to P . For any subset T of the Riemann space S , its holomorphy ring consists of all those $x \in K$ which are holomorphic at every place $P \in T$. By definition, this holomorphy ring is the intersection of the valuation rings $\mathcal{O}_P$ belonging to places $P \in T$. If T is empty then this intersection is to be interpreted as being the whole field K .

In the next theorem we will assume that K is formally p-adic over $\mathfrak{o}[u]$ and that k is p-adically closed. Then by Theorem 7.2 the basic subset S_u^z of S is non-empty. We will give an explicit description of the holomorphy ring of S_u^z . In the proof it will turn out to be convenient to consider also places of dimension n-1 on the function field $K|k$. Recall that n is the transcendence degree of K over k . To simplify notations let us collect in the set D_u^z all places P of dimension $n-1$ on $K|k$ such that

(i) $z_i P \neq \infty$ for $1 \leq i \leq s$, $u_j P \neq \infty$ for $1 \leq j \leq r$,

(ii) KP is formally p-adic over $\mathfrak{o}[uP]$.

It follows from the proof of Theorem 7.2 that $D_u^z \neq \emptyset$ if and only if $S_u^z \neq \emptyset$.

One more notation is needed. Assuming again that K is formally p-adic over $\mathfrak{o}[u]$, we conclude from Lemma 7.1 that $1 + \pi b \neq 0$ for all $b \in \mathcal{O}[u,\gamma K]$. We denote by R_u the subring of K consisting of elements which can be represented as

$$\frac{a}{1+\pi b} \quad \text{with} \quad a,b \in \mathcal{O}[u,\gamma K] .$$

This ring contains the γ-Kochen ring R_γ and thus by Corollary 6.16 is a Prüfer ring. Therefore R_u is equal to the intersection of all valuation rings above R_u which are centered at some maximal ideal of R_u . As in Theorem 6.8 , every maximal ideal of R_u contains π . Thus it follows from 6.8 that the valuation rings above R_u which are centered at some maximal ideal are precisely the p-valuation rings of p-rank d containing $\mathcal{O}[u]$. We will call R_u the <u>Kochen ring of</u> K <u>over</u> $\mathcal{O}[u]$. In case k is p-adically closed we know from Theorem 6.15 that $\mathcal{O} = \gamma k$. Hence we may replace $\mathcal{O}[u,\gamma K]$ by $\mathbb{Z}[u,\gamma K]$.

If K is not formally p-adic over $\mathcal{O}[u]$, we infer from Lemma 7.5 that π is a unit in $\mathcal{O}[u,\gamma K]$. Hence we have

$$1 + \pi\mathcal{O}[u,\gamma K] = \mathcal{O}[u,\gamma K] .$$

Using Merckel's Lemma (cf. Appendix) we get

$$K = k(\gamma K) = \mathrm{Quot}(\mathcal{O}[u,\gamma K]) .$$

Thus we can represent every element from K in the form

$$\frac{a}{1+\pi b} \quad \text{with} \quad a,b \in \mathcal{O}[u,\gamma K] \ , \ 1+\pi b \neq 0 .$$

Taking this again as the definition of R_u, we have $R_u = K$ in case K is not formally p-adic over $\mathcal{O}[u]$.

THEOREM 7.4 _Let_ $K|k$ _be a function field in_ n _variables over the p-adically closed field_ k . _Then the holomorphy ring of_ S_u^z _coincides with_ $R_u \cdot k[z]$.

Moreover, the holomorphy ring of S_u^z _is equal to the intersection of the valuation rings_ $\mathcal{O}_P$ _belonging to all_ $P \in D_u^z$.

By this theorem, a function $x \in K$ is holomorphic on a non-empty basic subset S_u^z if and only if it admits a representation of the form

$$x = \frac{t'}{1 + \pi t}$$

with $t' \in k[z,u,\gamma K]$ and $t \in \mathbb{Z}[u,\gamma K]$.

Proof: Clearly, we may assume that K is formally p-adic over $\mathfrak{o}[u]$. Hence S_u^z and D_u^z are both non-empty.

Let $P \in S_u^z$ or $P \in D_u^z$. In both cases KP is formally p-adic over $\mathfrak{o}[uP]$. The preimage of a p-valuation ring of p-rank d containing $\mathfrak{o}[uP]$ is a p-valuation ring $\mathcal{O}$ on K of p-rank d containing $\mathfrak{o}[u]$. Hence

$$R_u \subset \mathcal{O} \subset \mathcal{O}_P .$$

Since $z_i P \neq \infty$, also $k[z] \subset \mathcal{O}_P$. Therefore, the ring $R_u \cdot k[z]$ is contained in the intersection of the valuation rings $\mathcal{O}_P$ belonging to all places P from S_u^z . Similarly, it is contained in $\bigcap \mathcal{O}_P$ where P ranges over D_u^z .

Now consider some element $y \in K$, not in $R_u \cdot k[z]$. Then

y^{-1} is not a unit in $R_u \cdot k[z,y^{-1}]$. Indeed, if it would be a unit, we could find $m \in \mathbb{N}$ and $\lambda_o,\ldots,\lambda_m \in R_u \cdot k[z]$ such that

$$y = \lambda_o + \lambda_1 y^{-1} + \ldots + \lambda_m y^{-m} .$$

This implies that y is integral over $R_u \cdot k[z]$. As an extension of the Prüfer ring R_u this ring is again a Prüfer ring and hence integrally closed. Now $y \in R_u \cdot k[z]$ contradicts the assumption on y .

We now apply the next lemma to the non-unit y^{-1} of $R_u \cdot k[z,y^{-1}]$ and find some $P \in D_u^{z,y^{-1}}$ such that $y^{-1}P = 0$. Thus in particular $P \in D_u^z$ and $y \notin \mathcal{O}_P$. Applying now the Place Existence Theorem 7.2 to the function field $KP|k$, we find some P'' in the basic subset S_{uP}^{zP} of the Riemann space of $KP|k$. Let P' be the composition $P \circ P''$. Then obviously P' belongs to S_u^z and $y^{-1}P' = 0$, implying that $y \notin \mathcal{O}_{P'}$.

Altogether we have now proved that

$$R_u \cdot k[z] = \bigcap \mathcal{O}_P = \bigcap \mathcal{O}_{P'}$$

where P ranges over D_u^z and P' ranges over S_u^z .

q.e.d.

LEMMA 7.5 <u>Let</u> $K|k$ <u>be a function field in</u> n <u>variables over the p-adically closed field</u> k . <u>If</u> $x \neq 0$ <u>is a non-unit element of</u> $R_u \cdot k[z]$, <u>then there is</u> $P \in D_u^z$ <u>with</u> $xP = 0$.

<u>Proof</u>: Since x is a non-unit in the ring $R_u \cdot k[z]$, by general valuation theory there exists a place Q of $K|k$ such that $\mathcal{O}_Q$

contains $R_u \cdot k[z]$ and $xQ = 0$. Clearly, Q is trivial on k . Let $zQ = (z_1Q,\ldots,z_sQ)$ generate a subfield of KQ of transcendence degree t over k. Choose $x_1,\ldots,x_t \in K$ such that $x_1Q,\ldots,x_tQ$ are algebraically independent over k . The place Q is trivial on $k(x_1,\ldots,x_t)$. Since $xQ = 0$ and $x \neq 0$, also $x_1,\ldots,x_t,x_{t+1}=x$ are algebraically independent over k . We choose moreover $x_{t+2},\ldots,x_n \in K$ such that $x_1,\ldots,x_n$ form a transcendence base of $K|k$. We also choose $y \in K$ such that $K = k(x_1,\ldots,x_n,y)$ and there is some irreducible polynomial $f \in k[X_1,\ldots,X_n,Y]$, monic in Y , with $f(x_1,\ldots,x_n,y) = 0$. By this choice, $f_Y(x_1,\ldots,x_n,y) \neq 0$ and K is the field of quotients of $k[x_1,\ldots,x_n,y]$.

The valuation ring $\mathcal{O}_Q$ contains R_u . Hence we find a maximal ideal I above the center of $\mathcal{O}_Q$ in R_u . Then $\mathcal{O} = \mathrm{Quot}_I(R_u)$ is a p-valuation ring of p-rank d containing $\mathfrak{o}[u]$. Clearly, $\mathcal{O} \subset \mathcal{O}_Q$. Inside some p-adic closure L of K with respect to $\mathcal{O}$ let k' be the relative algebraic closure of $k(x_1,\ldots,x_{t+1})$ with induced p-valuation ring $\mathfrak{o}'$. By Theorem 3.4, k' is p-adically closed with respect to $\mathfrak{o}'$. Together with $\mathcal{O}$, the coarser valuation ring $\mathcal{O}_Q$ extends to L as a coarsening, and then also restricts to k' . If we denote by Q' the place of k' belonging to this restriction, we obtain $\mathfrak{o}' \subset \mathcal{O}_{Q'}$. Thus $\mathfrak{o}'Q'$ is a p-valuation ring on $k'Q'$ of p-rank d extending the unique p-valuation ring $\mathfrak{o}$ of k .

Since the p-adic closure L is henselian also with respect to $\mathcal{O}_Q$ and all z_iQ are algebraic over $k(x_1,\ldots,x_t)Q$, we can find elements $z_i' \in L$ algebraic over $k(x_1,\ldots,x_t)$ with $z_i'Q = z_iQ$.

From our construction of k' we know that $z_i' \in k'$. Observing $\mathcal{O} \subset \mathcal{O}_Q$ we thus find $z_i - z_i' \in \mathfrak{m}_Q \subset \pi\mathcal{O} \subset \mathcal{O}$.

Let us collect what we have obtained so far:

(1) $K\cdot k'|k'$ is a function field in $n-(t+1)$ variables, formally p-adic over $\mathcal{O}'[u,z-z']$,

(2) k' is p-adically closed with respect to $\mathcal{O}'$,

(3) Q' is a place of k', trivial on $k(x_1,\dots,x_t)$, such that $xQ' = 0$ and $\mathcal{O}'Q'$ is a p-valuation on $k'Q'$ of p-rank d.

We will now represent the elements z_i and u_j from K as quotients of functions from $k[x_1,\dots,x_n,y]$. Fixing such a representation we can talk about their denominators. Applying Theorem 7.2 suitably to the situation in (1), we can therefore find a rational place P' of $K\cdot k'|k'$ such that

$$\begin{array}{ll} x_\nu P' \neq \infty & \text{for all } 1 \leq \nu \leq n, \\ y P' \neq \infty \quad \text{and} \quad f_Y(x_1,\dots,x_n,y)P' \neq 0, & \\ \text{denominators of } z_i \text{ and } u_j \text{ have images } \neq 0, & \\ (z_i - z_i')P' \in \mathcal{O}' & \text{for all } 1 \leq i \leq s, \\ u_j P' \in \mathcal{O}' & \text{for all } 1 \leq j \leq r. \end{array}$$

Thus we have found elements $\bar{x}_1,\dots,\bar{x}_n,\bar{y} \in k'$ such that $\bar{x}_\nu = x_\nu$ for all $1 \leq \nu \leq t+1$, and

(i) $f(\bar{x}_1,\dots,\bar{x}_n,\bar{y}) = 0$, $f_Y(\bar{x}_1,\dots,\bar{x}_n,\bar{y}) \neq 0$,

(ii) denominators of $u_j(\bar{x}_1,\dots,\bar{y})$ and $z_i(\bar{x}_1,\dots,\bar{y})$ are different from 0,

(iii) $u_j(\bar{x}_1,\dots,\bar{y})$, $(z_i(\bar{x}_1,\dots,\bar{y})-z_i') \in \mathcal{O}'$.

We now use (i)-(iii) in order to construct the desired place of dimension $n-1$ on $K|k$. First we extend the place Q' from k' to a place P_1 of the field $K_1 = k'(x_{t+2},\dots,x_n)$ such that the residue field K_1P_1 is a purely transcendental extension of $k'Q'$ of transcendence degree $n-(t+1)$. Such extensions exist by general valuation theory (see e.g.[Ri], page 49). Next we extend $\mathcal{O}'Q'$ to some p-valuation ring of p-rank d on K_1P_1. This is possible by iterated application of Example 2.2. By Example 2.5 the inverse image $\mathcal{O}_1$ of this extended p-valuation on K_1P_1 is again a p-valuation ring of p-rank d. Finally let L_1 be some p-adic closure of K_1 with respect to $\mathcal{O}_1$. Since $\mathcal{O}_1 \subset \mathcal{O}_{P_1}$, the valuation ring $\mathcal{O}_{P_1}$ extends to a coarsening of the unique p-valuation ring on L_1. Let us keep the notations $\mathcal{O}_1$ and P_1 also for their extensions to L_1. Thus P_1 is a place of L_1 which extends the place Q' of k' such that $\mathcal{O}_1P_1$ is a p-valuation ring of p-rank d extending $\mathcal{O}'Q'$ from $k'Q'$ to the residue field L_1P_1 which has transcendence degree $n-(t+1)$ over $k'Q'$.

The final step of the construction will now be a suitable embedding of K into L_1 over k. With respect to the topology induced by $\mathcal{O}_1$ on L_1, the 'Implicite Function Theorem' holds in L_1 for polynomials. This is a consequence of the Henselian property of $\mathcal{O}_1$ (see [P-Z], Theorem 7.4). Therefore, if we choose elements $x_1^*,\dots,x_n^*$ in L_1 close enough to $\bar{x}_1,\dots,\bar{x}_n$, we can find $y^* \in L_1$ also very close to $\bar{y}$ such that $x_1^*,\dots,x_n^*,y^*$ satisfy properties (i)*,(ii)*, and (iii)* which are obtained from (i)-(iii) replacing $\bar{x}_1,\dots,\bar{x}_n,\bar{y}$ by $x_1^*,\dots,x_n^*,y^*$ and $\mathcal{O}'$ by $\mathcal{O}_1$. Our choice for

$x_1^*, \dots, x_n^*$ will be as follows. First we let

$$x_\nu^* = \bar{x}_\nu = x_\nu \qquad \text{for all } 1 \le \nu \le t+1 .$$

Next we continue by induction. If x_ν^* is already chosen for $t+1 \le \nu \le n$, we denote by k_ν' the relative algebraic closure of $k'(x_{t+2}^*, \dots, x_\nu^*)$ inside L_1. One easily checks that the elements of $L_1 \setminus k_\nu'$ come arbitrarily close to $\bar{x}_{\nu+1}$. Let $x_{\nu+1}^*$ be one of those elements which is close enough.

By the above choice, the elements $x_1^*, \dots, x_n^*$ are algebraically independent over k. Thus the prescriptions $x_\nu \mapsto x_\nu^*$ $(1 \le \nu \le n)$ and $y \mapsto y^*$ define a k-embedding of K into L_1. Denote the image by K^* and let P^* be the restriction of P_1 to K^*. Clearly, P^* is a place on $K^*|k$ of dimension $n-1$ with formally p-adic residue field. Moreover, we conclude from (ii)* and (iii)* that K^*P^* is even formally p-adic over $\mathcal{O}[u^*P^*]$ where we let $u_j^* = u_j(x_1^*, \dots, y^*)$. Putting also $z_i^* = z_i(x_1^*, \dots, y^*)$ which is possible by (ii)*, we conclude from (iii)* that

$$(z_i^* - z_i')P_1 \in \mathcal{O}_1 P_1 .$$

By the choice of $z_i' \in k'$ we know that

$$z_i'P_1 = z_i'Q' = z_iQ \neq \infty .$$

Hence also

$$z_i^*P^* = z_i^*P_1 \neq \infty .$$

If we finally omit all the stars (i.e. identify K with K^*) we have found a place P of $K|k$ of dimension $n-1$ such that

KP is formally p-adic over $\mathfrak{o}[uP]$, $z_iP \neq \infty$ for all $1 \leq i \leq s$, and $xP = 0$.

q.e.d.

Let us remark that Lemma 7.5 , formulated for rational places $P \in S_u^z$, was first proved in [J-R] using 'Zariski's Local Uniformization'. Later the use of this theorem was eliminated by F.V. Kuhlmann. In the above proof of Lemma 7.5 , the construction of k' and P' is similar to that of Kuhlmann.

7.3 Nullstellensatz and integral definite functions

We continue to use the notations of Subsections 7.1 and 7.2. In particular, $K|k$ is a function field in n variables. For fixed elements

$$z_i \quad (1 \leq i \leq s) \quad \text{and} \quad u_j \quad (1 \leq j \leq r)$$

of K , the basic subset S_u^z of the Riemann space S consists of those rational places P on $K|k$ such that

$$z_i P \neq \infty \ (1 \leq i \leq s) \quad \text{and} \quad u_j P \in \mathfrak{o} \quad (1 \leq j \leq r)$$

where $\mathfrak{o}$ denotes the fixed p-valuation ring of p-rank d on k . Recall also that K is called formally p-adic over $\mathfrak{o}[u]$ if there exists a p-valuation of p-rank d on K containing $\mathfrak{o}[u]$. In case k is p-adically closed, by Theorem 7.2 , K is formally p-adic over $\mathfrak{o}[u]$ if and only if S_u^z is non-empty.

THEOREM 7.6 (_Nullstellensatz_) _In the situation described above assume that_ k _is p-adically closed and_ $S_u^z \neq \emptyset$. _Let_ $g, f_1, \dots, f_m \in K$ _be holomorphic functions on_ S_u^z .

A necessary and sufficient condition for g _to vanish at all common zeros_ $P \in S_u^z$ _of_ $f_1, \dots, f_m$ (i.e. $f_1 P = \dots = f_m P = 0$ _implies_ $gP = 0$ _for all_ $P \in S_u^z$) _is that some power_ g^N _admits a representation_

$$g^N = \lambda_1 f_1 + \dots + \lambda_m f_m$$

where the $\lambda_\nu \in K$ _are holomorphic functions on_ S_u^z .

Recall that by Theorem 7.4 the ring of holomorphic functions on S_u^z is equal to $R_u \cdot k[z]$. To simplify notations let us write H_u^z for this ring.

Proof: By Theorem 6.17 the Kochen ring R_Y is a Bezout ring. Hence the overring H_u^z is a Bezout ring too. Therefore it suffices to consider only the case $m = 1$. Thus we are assuming that

$$fP = 0 \text{ implies } gP = 0 \text{ for all } P \in S_u^z$$

where f and g are holomorphic functions on S_u^z . From this assumption we see that f has no zero on $S_u^{z,g^{-1}}$. Thus by Lemma 7.5 (and the proof of 7.4), f is a unit in $H_u^{z,g^{-1}}$. Observing that

$$H_u^{z,g^{-1}} = R_u \cdot k[z, g^{-1}] = H_u^z[g^{-1}]$$

we can find a representation of f^{-1} of the form

$$f^{-1} = \lambda_o + \lambda_1 g^{-1} + \dots + \lambda_N g^{-N}$$

with $N \in \mathbb{N}$ and $\lambda_o,\ldots,\lambda_N \in H_u^z$. Clearly, this implies $g^N \in H_u^z \cdot f$.

q.e.d.

Let us now call a function $f \in K$ integral definite on S_u^z if $fP \in \mathfrak{o}$ for all $P \in S_u^z$.

THEOREM 7.7 Assume that k is p-adically closed and $S_u^z \neq \emptyset$. Then a function $f \in K$ is integral definite on S_u^z if and only if f belongs to R_u, i.e. f admits a representation of the form

$$f = \frac{a}{1+\pi b} \quad \text{with} \quad a,b \in \mathfrak{z}[u,\gamma K].$$

Proof: Let us first assume that $f \in R_u$. For a fixed place $P \in S_u^z$ denote by $\mathcal{O}$ the preimage of $\mathfrak{o}$ under P. From the assumptions on u and $\mathfrak{o}$ we have

$$\mathfrak{z}[u,\gamma K] = \mathfrak{o}[u,\gamma K] \subset \mathcal{O}.$$

Thus f belongs to $\mathcal{O}$ too, proving that fP belongs to $\mathfrak{o}$.

Conversely, assume that $fP \in \mathfrak{o}$ for all $P \in S_u^z$. If f is not in R_u there exists a p-valuation ring $\mathcal{O}$ of p-rank d over $\mathfrak{o}[u]$ such that $f \notin \mathcal{O}$. Hence $g = (\pi f)^{-1} \in \mathcal{O}$, proving that K is formally p-adic over $\mathfrak{o}[u,g]$. Hence by Theorem 7.4, $S_{u,g}^z$ is non-empty. Let $P \in S_{u,g}^z$. Then

$$\frac{1}{\pi} \cdot \frac{1}{fP} = gP \in \mathfrak{o}$$

implies the contradiction $fP \notin \mathfrak{o}$.

q.e.d.

We now turn to the geometric interpretation of the above theorems. In doing so, $\mathfrak{o}$ still is a fixed p-valuation ring on k of p-rank d.

In the following V always denotes some affine variety defined over k. By $V(k)$ we denote the set of k-rational points $a = (a_1,\dots,a_n)$ on V. Let $x = (x_1,\dots,x_n)$ denote some generic point of V over k. Then $k[x]$ is the coordinate ring and $k(x) = K$ the function field of V over k. Let now

$$u = (u_1,\dots,u_r)$$

be a family of elements from $k[x]$. As above R_u denotes the Kochen ring of K over $\mathfrak{o}[u]$. By $V_u(k)$ we denote the set of k-rational points $a \in V(k)$ such that

$$u_j(a) \in \mathfrak{o} \quad \text{for all} \quad 1 \leq j \leq r .$$

THEOREM 7.8 _Let_ V _be an affine variety, defined over the_ p-_adically closed field_ k.

_If_ $V_u(k)$ _contains a simple point, then the function field_ K _of_ V _is formally p-adic over_ $\mathfrak{o}[u]$.

_If_ K _is formally p-adic over_ $\mathfrak{o}[u]$, _then the set of simple points of_ $V_u(k)$ _is non-empty and Zariski-dense in_ $V(k)$.

Thus in particular, there is a simple k-rational point on V if and only if K is formally p-adic over $\mathfrak{o}$. In case $k = \mathbb{Q}_p$ this is exactly Theorem 2_p from the introduction.

Proof: First let $a = (a_1,\dots,a_n) \in V_u(k)$ be a simple point. As it is well-known, there exists a rational place of $K|k$ such that

$$x_i P = a_i \qquad (1 \le i \le n) .$$

From the assumption $u_j(a) \in \mathfrak{o}$ we therefore get $u_j P \in \mathfrak{o}$ $(1 \le j \le r)$. Thus pulling $\mathfrak{o}$ back through P we find a p-valuation ring on K of p-rank d containing $\mathfrak{o}[u]$.

Conversely, assume that K is formally p-adic over $\mathfrak{o}[u]$. Then by Theorem 7.2 , S_u^z is non-empty. We are free to choose $z = (z_1,\dots,z_s)$. First we require that z contains all x_i $(1 \le i \le n)$. Moreover, if $f \in k[x]$ defines a hypersurface containing all singular points of V and $g \in k[x]$ is arbitrary non-zero, we require z to contain $(f\cdot g)^{-1}$. Now let $P \in S_u^z$. Then the point a defined by

$$a_i = x_i P \qquad (1 \le i \le n)$$

is a simple point of $V(k)$ satisfying $g(a) \neq 0$ and $u_j(a) \in \mathfrak{o}$. Since g was arbitrary we thus obtain Zariski density. q.e.d.

THEOREM 7.9 _Let_ V _be an affine variety, defined over the p-adically closed field_ k . _Assume that_ $g, f_1,\dots,f_m \in k[x]$ _and_ $V_u(k)$ _contains a simple point of_ V .

_If_ g _vanishes at all common zeros of_ $f_1,\dots,f_m$ _in_ $V_u(k)$ _then some power_ g^N _admits a representation_

$$g^N = \lambda_1 f_1 + \dots + \lambda_m f_m \quad \text{with} \quad \lambda_i \in R_u\cdot k[x] .$$

<u>Conversely, if this condition is satisfied, then</u> g <u>vanishes at all zeros of</u> $f_1,\dots,f_m$ <u>in</u> $V_u(k)$ <u>which are simple points of</u> V .

Consequently, if the variety V is non-singular, then the above condition is necessary and sufficient for g to vanish at all common zeros of $f_1,\dots,f_m$ in $V_u(k)$.

<u>Proof</u>: Let $P \in S_u^x$. Then $a = xP$ is a point on $V_u(k)$. Thus by the assumption of the first claim,

$$f_1P = \dots = f_mP = 0$$

implies $gP = 0$. Since this holds for all $P \in S_u^x$, the assertion on g follows from Theorem 7.6 .

Conversely, let g satisfy the above condition. Then again by Theorem 7.6,

$$f_1P = \dots = f_mP = 0$$

implies $gP = 0$ for all $P \in S_u^x$. Thus, if $a \in V_u(k)$ is a simple point of V and P is a rational place of $K|k$ with $a = xP$, then we have $P \in S_u^x$, and therefore

$$f_1(a) = \dots = f_m(a) = 0$$

implies $g(a) = 0$.

q.e.d.

From this very general 'Nullstellensatz' we can easily deduce the following special case.

COROLLARY 7.10 <u>Let</u> k <u>be p-adically closed and</u> $f_1,\dots,f_m$, $g \in k[X_1,\dots,X_n]$. <u>A necessary and sufficient condition for</u> g <u>to</u>

vanish at all common zeros of $f_1,\dots,f_m$ in k^n is that some power g^N admits a representation

$$g^N = \lambda_1 f_1 + \dots + \lambda_m f_m$$

with $\lambda_i \in R\cdot k[X_1,\dots,X_n]$ $(1 \le i \le m)$.

Proof: In this case V is the affine n-space and u is empty. That $k(X_1,\dots,X_n)$ is formally p-adic over $\mathfrak{o}$ can be seen from Example 2.2 .

q.e.d.

In the above corollary, the λ_i admit a representation

$$\lambda_i = \frac{t'_i}{1+\pi t_i}$$

with $t'_i \in k[X,\gamma k(X)]$ and $t_i \in \mathbb{Z}[\gamma k(X)]$. If we consider only zeros from $\mathfrak{o}^n$ instead of k^n in Corollary 7.10, then by taking $u = (X_1,\dots,X_n)$, we get

$$t'_i \in k[X,\gamma k(X)] \quad \text{and} \quad t_i \in \mathbb{Z}[X,\gamma k(X)] .$$

Let us now turn to the 'p-adic Analog of Hilbert's 17th Problem'. It is easy to obtain a translation of Theorem 7.7 into the language of varieties, similar to the translation of the Nullstellensatz. We leave it to the reader to treat the general case. Here, we will only deal with the affine n-space. The following theorem may be considered as the 'p-adic Analog of Hilbert's 17th Problem'.

THEOREM 7.11 Let k be p-adically closed. If $f,g \in k[X_1,\ldots,X_n]$ and f/g is integral definite (i.e. $g(a) \neq 0$ implies $f(a)/g(a) \in \mathfrak{o}$ for all $a \in k^n$), then there are $t',t \in \mathbb{Z}[\gamma k(X_1,\ldots,X_n)]$ such that

$$\frac{f}{g} = \frac{t'}{1+\pi t} \quad .$$

Proof: In Theorem 7.7 we let $K = k(X_1,\ldots,X_n)$, $u = \emptyset$, and $z = (X_1,\ldots,X_n,g^{-1})$. Clearly, S^z is non-empty. For every $P \in S^z$ we get

$$g(X_1P,\ldots,X_nP) = gP \neq 0 \quad .$$

Hence from the assumption we infer

$$(f/g)P \in \mathfrak{o} \quad .$$

Now Theorem 7.7 yields the desired representation for f/g.

q.e.d.

We finish this section by proving a theorem in the p-adic case which parallels a theorem of McKenna in the real case (see [MK]). Before doing so, let us consider the following situation. Assume k carries a p-valuation v and is dense in some p-adic closure F with respect to v. From the density one easily deduces that v extends immediate to F. Thus in particular, the group of values does not change by passing from k to F. Therefore, by Theorem 3.1, vk is a $\mathbb{Z}$-group, proving that k admits (up to isomorphism) only one p-adic closure (cf. Theorem 3.2). By this observation, the phrase "k is dense in its p-adic closure with respect to v'' is unambiguous.

THEOREM 7.12 *If* k *is dense in its p-adic closure with respect to* $\mathcal{O}$*, then every integral definite function from* $k(X_1,\dots,X_n)$ *can be represented as*

$$\frac{t'}{1+\pi t} \quad \text{with} \quad t',t \in \mathcal{O}[\gamma k(X_1,\dots,X_n)] .$$

Conversely, if such a representation exists for all integral definite functions from $k(X_1)$*, then* k *is dense in its p-adic closure with respect to* $\mathcal{O}$.

Proof: Assume first that $f,g \in k[X_1,\dots,X_n]$ and f/g is integral definite on k^n . If f/g does not allow a representation of the desired form, by Theorem 6.13 there is a p-valuation ring $\mathcal{O}$ on $K = k(X_1,\dots,X_n)$ of p-rank d such that $f/g \notin \mathcal{O}$. Let F be the relative algebraic closure of k inside some p-adic closure of K with respect to $\mathcal{O}$. By Theorem 3.4, F is also p-adically closed. Denote the unique p-valuation ring of F again by $\mathcal{O}$. Since the function f/g is not in $\mathcal{O}$, it also cannot belong to the Kochen ring of $F(X_1,\dots,X_n)$. Thus by Theorem 7.11 we can find some $a \in F^n$ such that $g(a) \neq 0$ and $f(a)/g(a) \notin \mathcal{O}$. From the density assumption and the continuity of polynomials we then deduce the existence of some $a' \in k^n$ such that $g(a') \neq 0$ and $f(a')/g(a') \notin \mathcal{O}$, which is a contradiction.

For the converse we assume that every integral definite function of $K = k(X_1)$ belongs to the Kochen ring of K over $\mathcal{O}$. We let $\hat{k}$ denote the completion of k with respect to the uniformity induced by the valuation ring $\mathcal{O}$. This completion is again a valued field with closure $\hat{\mathfrak{o}}$ of $\mathfrak{o}$ as an immediate extension of $\mathfrak{o}$. Hence $\hat{\mathfrak{o}}$ is again a p-valuation ring of p-rank d . Clearly, k is dense in $\hat{k}$. Supposing now that k is not dense in any p-adic closure, the relative algebraic closure k' of k inside $\hat{k}$ is not p-adically closed. Let F be some p-adic closure of k' with respect to $\hat{\mathfrak{o}} \cap k'$. There exists $x \in F \setminus k'$. Let $f(X_1)$ be the irreducible polynomial of x over k . The set f(k) cannot approach zero, since otherwise f would have a zero in $\hat{k}$ and hence in k' (see [Ka], Theorem 1) . After multiplying f with some suitable constant from k and calling the result again f , we may therefore assume that

$$\pi\mathfrak{o} \cap f(k) = \emptyset \quad .$$

Thus the function f^{-1} is integral definite on k . By assumption, f^{-1} then admits a representation

$$f^{-1} = \frac{t'}{1 + \pi t}$$

with $t', t \in \mathfrak{o}[\Upsilon k(\underline{X})]$. Passing to F , the right hand side is still integral for all substitutions. This contradicts the fact that f has a zero in F .

q.e.d.

Since the field $\mathbb{Q}$ of rational numbers is dense in $\mathbb{Q}_p$ and hence in the p-adic closure of $\mathbb{Q}$ with respect to its unique p-valuation, we may apply Theorem 7.12 to this situation. Therefore every integral definite function $f \in \mathbb{Q}(X_1,\ldots,X_n)$ can be represented as

$$f = \frac{t'}{1 + pt}$$

with $t, t' \in \mathbb{Z}[\Upsilon\mathbb{Q}(X_1,\ldots,X_n)]$. Here we also use the fact that the unique p-valuation ring of $\mathbb{Q}$ is equal to the Kochen ring of $\mathbb{Q}$. This is an immediate consequence of Theorem 6.14 .

Appendix

The following lemma has been used in §6 for the Kochen operator γ . The lemma itself, however, is of general nature and does not refer to the theory of formally p-adic fields. We consider the following situation:

K an arbitrary field

F an extension field of K

$\gamma(X)$ a rational function in $K(X)$ such that its differential $d\gamma(X) \neq 0$.

The condition $d\gamma(X) \neq 0$ is equivalent to saying that $\gamma(X)$ should not be constant and, if the characteristic of K is a prime number $p > 0$, then $\gamma(X)$ should not be a function of X^p . Obviously this condition is satisfied if $\gamma(X)$ is the Kochen operator as defined in the text above. Again in the general case, let γF denote the set of elements $\gamma(f)$ where $f \in F$ and $\gamma(f) \neq \infty$, i.e. $\gamma(f)$ should be defined as an element in F . The field $K(\gamma F)$ generated by γF over K , is then a subfield of F .

We choose $f(X), g(X) \in K[X]$ such that

$$\gamma(X) = \frac{f(X)}{g(X)}$$

and f and g are relatively prime. Let M denote the maximal degree of f and g . We now can state

THEOREM (Merckel's Lemma) *If the base field* K *has at least* $(M+1)^2$ *elements, then* $K(\gamma F) = F$.

Proof: We distinguish the two cases where K is infinite or finite.

Case 1: K is infinite (cf. [J-R], Appendix B)

(i) First we consider the case that F is a rational function field in one variable:

$$F = K(x) .$$

For brevity let us put

$$F_o = K(\gamma F) .$$

We have to show that $F_o = F$. The field F_o contains the non-constant rational function $\gamma(x)$; hence F_o is transcendental over K. It follows that F is algebraic and of finite degree over F_o. Moreover F is separable over F_o, in view of our hypothesis that the differential $d\gamma(x)$ does not vanish. It follows that there are only finitely many places P of $F|K$ which are ramified over F_o (these places need not be rational over K).

Now let us consider the automorphism group G of $F|K$. It is well known that every $\sigma \in G$ can be represented in the form

$$x\sigma = \frac{ax+b}{cx+d}$$

with coefficients $a, b, c, d \in K$ and nonvanishing determinant. (We write σ as right operator.) For any $f = f(x) \in F$ we have

$$f\sigma = f\left(\frac{ax+b}{cx+d}\right) .$$

Since $\gamma(f)\cdot\sigma = \gamma(f\sigma)$ we conclude

$$(\gamma F)\cdot\sigma = \gamma F$$

$$F_o\sigma = F_o \ .$$

Thus every automorphism $\sigma \in G$ maps the field F_o onto itself. Consequently G permutes the finitely many places of F which are ramified over F_o .

Recall that G acts naturally on the places P of $F|K$; the image σP of P is given by the formula

$$x\cdot\sigma P = x\sigma\cdot P \ .$$

We do not assume that P is K-rational; thus P is an arbitrary place of F over K with values in the algebraic closure $\tilde{K}$ of K. In the above formula, the values $x\cdot\sigma P$ and $x\sigma\cdot P$ are understood to be elements in $\tilde{K} \cup \infty$.

Let G_o denote the normal subgroup of G which leaves every ramified place of $F|F_o$ fixed. By what has been said above G_o is of finite index in G . It follows that G_o contains infinitely many translations τ of the form

$$x\tau = x + b$$

with $b \in K$. Notice that the field K is supposed to be infinite; hence indeed the group T of all translations $\tau \in G$ is infinite and thus $G_o \cap T$ is infinite too. Let $\tau \in G_o \cap T$, $\tau \neq 1$. Then $b \neq 0$. If the place P of $F|K$ is ramified over F_o then $\tau P = P$ and hence

$$xP = x\cdot\tau P = x\tau\cdot P = (x+b)\cdot P = xP + b \ .$$

Since $b \neq 0$ we conclude $xP = \infty$. Hence there is at most one place P of $F|K$ which is ramified over F_o, namely the pole of x. After replacing x by x^{-1} the pole of x becomes the zero of x^{-1}; hence the pole of x is not ramified either. In other words: F is unramified over F_o.

From Lüroth's Theorem we know that F_o is a rational function field over K. Now a rational function field does not admit any proper separable-algebraic field extension which is unramified and preserves the field of constants; this is well known from the general ramification theory of function fields. We conclude $F = F_o$, as contended.

(ii) Now let F be an arbitrary extension field of K. Let X be an indeterminate over F. In (i) we have proved that $X \in K(\gamma K(X))$. This means that there is a relation of the form

$$(*) \qquad X = \Phi(\gamma(f_1(X)),\ldots,\gamma(f_n(X)))$$

where Φ denotes a rational function in n variables with coefficients in K, and where $f_1,\ldots,f_n \in K(X)$. Let $a \in F$ be such that all rational functions involved on the right hand side of (*) are defined at a, and that the specialization $X \rightarrow a$ yields the relation

$$(**) \qquad a = \Phi(\gamma(f_1(a)),\ldots,\gamma(f_n(a))).$$

This condition excludes only a finite number of elements in F. For all remaining $a \in F$ we infer from (**) that $a \in K(\gamma K(a)) \subset K(\gamma K)$. Thus we have seen that $K(\gamma F)$ contains all but finitely many elements $a \in F$. Since F is infinite this implies in fact that $K(\gamma F) = F$.

<u>Case 2</u>: Let us again put

$$F_o = K(\gamma F)$$

and assume that $F_o \neq F$. Hence $[F:F_o] \geq 2$ which implies

$$|F| \geq |F_o|^2 .$$

Let $x \in F$ be such that $\gamma(x)$ is defined. If we put $\gamma(x) = a \in F_o$, then x is a zero of the polynomial

$$\varphi_a(X) = f(X) - a \cdot g(X) \in F_o[X].$$

For each $a \in \gamma(F)$ denote by C_a the set of zeros $x \in F$ of $\varphi_a(X)$, i.e. the set of $x \in F$ such that $\gamma(x) = a$. Moreover, let C_∞ denote the set of those $x \in F$ such that $g(x) = 0$. We then obtain

$$|F| = \sum_{a \in \gamma F} |C_a| + |C_\infty| \leq |\gamma F| \cdot M + \deg g .$$

Using the above obtained inequality, we find

$$|F| \leq |F_o| \cdot M + \deg g \leq |F|^{\frac{1}{2}} \cdot M + \deg g$$

$$|F|^{\frac{1}{2}} \leq \frac{1}{2} (M + \sqrt{M^2 + 4\deg g})$$

$$(+) \qquad |F| \leq \frac{1}{2} M(M + \sqrt{M^2 + 4\deg g}) + \deg g .$$

Since $\deg g < M$ inequality (+) implies

$$|F| \leq \frac{1}{2} \cdot M(M + \sqrt{M^2 + 4M}) + M \leq M^2 + 2M .$$

Therefore we get a contradiction to the assumption $F_o \neq F$ whenever we have

$$|F| > M(M+2)$$

or equivalently

$$|F| \geq M(M+2) + 1 = (M+1)^2.$$

q.e.d.

Let us remark that in case $\gamma(X)$ is a polynomial (and hence $\deg g = 0$) we can derive from (+) the inequality

$$|F| \leq M^2 = (\deg f)^2.$$

Now $K(\gamma F) = F$ already holds if F has more than M^2 elements.

References

[A] E. Artin: Über die Zerlegung definiter Funktionen in Qudrate, Abh. Math. Sem. Univ. Hamburg 5 (1927), 100-115.

[A-S] E.Artin-O.Schreier: Eine Kennzeichnung der reell-abgeschlossenen Körper, Abh. Math. Sem. Univ. Hamburg 5(1927),225-231.

Algebraische Konstruktion reeller Körper, Anhandlungen Math. Sem. Univ. Hamburg 5(1927), 85-99.

[Ax] J. Ax: A metamathematical approach to some problems in number theory, Appendix, AMS Proc. Symp.Pure Math.XX(1971),161-190.

[A-K] J.Ax-S.Kochen: Diophantine problems over local fields
I Amer. J. Math. 87 (1965), 605-630;
II Amer. J. Math. 87 (1965), 631-648;
III Ann. of Math. 83 (1966), 437-456.

[B-S] J.L.Bell-A.B.Slomson: Models and ultraproducts: an introduction, North Holland 1969.

[C-K] C.C.Chang-H.J.Keisler: Model theory, North Holland 1973.

[E] O. Endler: Valuation theory, Universitext, Springer 1972.

[Er] Ju.L.Eršov: On elementary theories of local fields, Algebra i logika Sem.4 (1965), no. 2, 5-30.

[J-R] M.Jarden-P.Roquette: The Nullstellensatz over $\mathcal{G}$-adically closed fields, J.Math.Soc.Japan 32(1980), 425-460.

[Ka] I.Kaplansky: Polynomials in topological fields, Bull.AMS 54 (1948), 909-916.

[K] S.Kochen: Integer valued rational functions over the p-adic numbers. A p-adic analogue of the theory of real fields. In: AMS Proc. Symp. Pure Math. XII (1969), 57-73.

[MI] A.MacIntyre: On definable subsets of p-adic fields, Journ. of Symbolic Logic 41 (1976), 605-611.

[MK] K.McKenna: New facts about Hilbert's seventeenth problem, Lecture Notes in Math.498, Springer 1975, 220-230.

[N] W. Narkiewicz: Elementary and analytic theory of algebraic numbers, Warszawa 1974.

[P_1] A. Prestel: Lectures on formally real fields, Monografias de Matemática 22, IMPA, Rio de Janeiro 1975.

[P_2] A. Prestel: Sums of squares over fields, Atas da 5.escola de álgebra, IMPA, Rio de Janeiro 1978, 33-44.

[P-Z] A.Prestel-M.Ziegler: Model theoretic methods in the theory of topological fields, J.reine angew.Math.299/300(1978),318-341.

[Ri] P.Ribenboim: Théorie des valuations, 2e edition, Les Presses de l'Université de Montréal 1968.

[R] P. Roquette: Principal ideal theorems for holomorphy rings in fields, J. reine angew. Math. 262/263(1973), 361-374.

[S] G.Sacks: Saturated model theory, Benjamin, Reading (Mass.) 1972.

[Z-S] O.Zariski-P.Samuel: Commutative algebra II, van Nostrand 1960

Notation index

Subject index